Basic Mechanical Engineering

Basic Mechanical Engineering

Brayden Anderson

NY RESEARCH PRESS
New York

Published by NY Research Press
118-35 Queens Blvd., Suite 400,
Forest Hills, NY 11375, USA
www.nyresearchpress.com

Basic Mechanical Engineering
Brayden Anderson

International Standard Book Number: 978-1-63238-855-1 (Hardback)

Cataloging-in-Publication Data

Basic mechanical engineering / Brayden Anderson.
p. cm.
Includes bibliographical references and index.
ISBN 978-1-63238-855-1
1. Mechanical engineering. 2. Machinery. 3. Engineering. I. Anderson, Brayden.
TJ145 .B37 2022
621--dc23

Contents

Preface

The application of physics, engineering mathematics, materials science and engineering for designing, analyzing, maintaining and manufacturing mechanical systems is known as mechanical engineering. It makes use of core principles from a large number of areas such as structural analysis, mechanics, dynamics and thermodynamics. Structural analysis is devoted towards analyzing the reasons behind the failure of objects and fixing them. Mechanics studies the forces as well as the effect they have on matter. Mechanical engineering is used to analyze and design heating and cooling systems, industrial equipment, transport systems, aircraft, watercraft, medical devices, etc. Its principles are applied in numerous fields such as robotics, bioengineering, tribology and nuclear engineering. This book discusses the fundamentals as well as modern approaches of mechanical engineering. The topics covered herein offer the readers new insights in this field. This book is appropriate for students seeking detailed information in this area as well as for experts.

A short introduction to every chapter is written below to provide an overview of the content of the book:

Chapter 1 - The branch of engineering which applies principles of physics, engineering mathematics and materials science for the design, analysis and manufacturing of mechanical systems is known sa mechanical engineering. This chapter has been carefully written to provide an easy understanding of the varied facets of mechanical engineering as well as the related fields.; **Chapter 2** - The branch of science which deals with the behavior of physical bodies when they are subjected to forces or displacements along with the effects which these bodies have on the environment is known as mechanics. This chapter discusses in detail the concepts related to mechanics such as motion, force and friction.; **Chapter 3** - The ability of withstanding an applied load without deformation or failure is referred to as strength of material. Some of the key areas of study related to the strength of materials are stress, strain and stress-strain curve. This chapter closely examines these important concepts related to strength of materials to provide an extensive understanding of the subject.; **Chapter 4** - Fluid mechanics is the domain of physics which studies the mechanics of different fluids such as liquids, plasma and gases, as well as the forces which act on them. There are a number of hydraulic machines like hydraulic turbines and pumps which are designed using fluid mechanics. The topics elaborated in this chapter will help in gaining a better perspective about the diverse concepts related to fluid mechanics and hydraulic machines.; **Chapter 5** - The process of converting resources or energy into useful mechanical forms in order to obtain desired output is known as machine design. The machines which are used to shape metals and other rigid materials are known as machine tools. This chapter has been carefully written to provide an easy understanding of the varied facets of machine design and machine tools as well as the manufacturing processes associated with them.

Finally, I would like to thank my fellow scholars who gave constructive feedback and my family members who supported me at every step.

Brayden Anderson

1

Understanding Mechanical Engineering

The branch of engineering which applies principles of physics, engineering mathematics and materials science for the design, analysis and manufacturing of mechanical systems is known as mechanical engineering. This chapter has been carefully written to provide an easy understanding of the varied facets of mechanical engineering as well as the related fields.

Mechanical engineering is a discipline of engineering that applies the principles of physics and materials science for analysis, design, manufacturing, and maintenance of mechanical systems. It is the branch of engineering that involves the production and usage of heat and mechanical power for the design, production, and operation of machines and tools. It is one of the oldest and broadest engineering disciplines.

The engineering field requires an understanding of core concepts including mechanics, kinematics, thermodynamics, materials science, and structural analysis. Mechanical engineers use these core principles along with tools like computer-aided engineering and product lifecycle management to design and analyze manufacturing plants, industrial equipment and machinery, heating and cooling systems, transport systems, aircraft, watercraft, robotics, medical devices and more.

Mechanical Engineering science emerged in the 19th century as a result of developments in the field of physics. The field has continually evolved to incorporate advancements in technology. Mechanical engineers today are pursuing developments in fields such as composites, mechatronics and nano technology. Mechanical Engineering overlaps with aerospace engineering, civil engineering, electrical engineering, petroleum engineering and chemical engineering to varying amounts.

Mechanical Engineering Functions

Four functions of the mechanical engineer, common to all branches of mechanical engineering, can be cited. The first is the understanding of and dealing with the bases of mechanical science. These include dynamics, concerning the relation between forces and motion, such as in vibration; automatic control; thermodynamics, dealing with the relations among the various forms of heat, energy, and power; fluid flow; heat transfer; lubrication; and properties of materials.

Second is the sequence of research, design, and development. This function attempts to bring about the changes necessary to meet present and future needs. Such work requires a clear understanding of mechanical science, an ability to analyze a complex system into its basic factors, and the originality to synthesize and invent.

Third is production of products and power, which embraces planning, operation, and maintenance. The goal is to produce the maximum value with the minimum investment and cost while maintaining or enhancing longer term viability and reputation of the enterprise or the institution.

Fourth is the coordinating the function of the mechanical engineer, including management, consulting and in some cases, marketing.

In these functions there is a long continuing trend toward the use of scientific instead of traditional or intuitive methods. Operations research, value engineering, and PABLA (problem analysis by logical approach) are typical titles of such rationalized approaches. Creativity, however, cannot be rationalized. The ability to take the important and unexpected step that opens up new solutions remains in mechanical engineering, as elsewhere, largely a personal and spontaneous characteristic.

Development of Machines for the Production of Goods

The high standard of living in the developed countries owes much to mechanical engineering. The mechanical engineer invents machines to produce goods and develops machine tools of increasing accuracy and complexity to build the machines.

The principal lines of development of machinery have been an increase in the speed of operation to obtain high rates of production, improvement in accuracy to obtain quality and economy in the product, and minimization of operating costs. These three requirements have led to the evolution of complex control systems.

The most successful production machinery is that in which the mechanical design of the machine is closely integrated with the control system. A modern transfer (conveyor) line for the manufacture of automobile engines is a good example of the mechanization of a complex series of manufacturing processes. Developments are in hand to automate production machinery further, using computers to store and process the vast amount of data required for manufacturing a variety of components with a small number of versatile machine tools.

Development of Machines for the Production of Power

The steam engine provided the first practical means of generating power from heat to augment the old sources of power from muscle, wind, and water. One of the first challenges to the new profession of mechanical engineering was to increase thermal efficiencies and power; this was done principally by the development of the steam turbine and associated large steam boilers. The 20th century has witnessed a continued rapid growth in the power output of turbines for driving electric generators, together with a steady increase in thermal efficiency and reduction in capital cost per kilowatt of large power stations. Finally, mechanical engineers acquired the resource of nuclear energy, whose application has demanded an exceptional standard of reliability and safety involving the solution of entirely new problems.

The mechanical engineer is also responsible for the much smaller internal combustion engines, both reciprocating (gasoline and diesel) and rotary (gas-turbine and Wankel) engines, with their widespread transport applications. In the transportation field generally, in air and space as well as on land and sea, the mechanical engineer has created the equipment and the power plant, collaborating increasingly with the electrical engineer, especially in the development of suitable control systems.

Development of Military Weapons

The skills applied to war by the mechanical engineer are similar to those required in civilian applications, though the purpose is to enhance destructive power rather than to raise creative efficiency. The demands of war have channeled huge resources into technical fields, however, and led to developments that have profound benefits in peace. Jet aircraft and nuclear reactors are notable examples.

Environmental Control

The earliest efforts of mechanical engineers were aimed at controlling the human environment by draining and irrigating land and by ventilating mines. Refrigeration and air conditioning are examples of the use of modern mechanical devices to control the environment.

Many of the products of mechanical engineering, together with technological developments in other fields, give rise to noise, the pollution of water and air, and the dereliction of land and scenery. The rate of production, both of goods and power, is rising so rapidly that regeneration by natural forces can no longer keep pace. A rapidly growing field for mechanical engineers and others is environmental control, comprising the development of machines and processes that will produce fewer pollutants and of new equipment and techniques that can reduce or remove the pollution already generated.

Mechatronics

Mechatronics is a branch of engineering that focuses on designing, manufacturing and maintaining products that have both mechanical and electronic components. The term was coined in 1969 by engineer Tetsuro Mori to describe the synergy that exists between electrical control systems and the mechanical machines they regulate. Since then, the meaning of the term has broadened to include the integration of multiple other disciplines, including computer engineering, systems engineering and programming.

As a design philosophy, mechatronics values systems thinking and an interdisciplinary approach to problem solving. Fairly recently in information technology (IT), the devops movement broke down the wall between software development and operations, encouraging IT pros to work in cross-departmental teams. Mechatronics has caused the same sea change in engineering. In many industries today, engineers have to work in cross-discipline collaborative teams in order to ensure that the complex, highly integrated systems they are designing will run smoothly.

To understand the concept of mechatronics, one can simply think about today's automobiles. The average car today has between 25 and 50 central processing units (CPUs) that control mechanical functions. A driver support system (DSS) such as anti-lock broke is designed with mechatronics; the electronic control system takes over the braking function when sensors recognize that one or more wheels are locking up. A pneumatic tire pressure monitoring system is also designed with mechatronics; each tire has a sensor inside that sends data to an on board electronic control system. If the pressure on one tire is low, the embedded software in the control system sends an alert to the vehicle's dashboard and a tire gauge icon lights up. Even an automobile's air bag is designed with mechatronics; a micro-electrical machine (MEM) in the front of the automobile will deploy an airbag when sensor data indicates rapid deceleration.

Thinking holistically has allowed engineers to make great strides in advancing automation and robotics. It has also allowed them to create smarter products, smaller products and tiny, efficient product components at the nanotech level. In order to successfully operate, diagnose and maintain electromechanical, automated equipment, it's become necessary for support technicians to re-skill formally and informally on a frequent basis.

Robotics

The term robotics refers to the technology that involves the design, manufacture, and uses of robots. Someone who works in the field of robotics is called a roboticist. This person requires a working knowledge of electronics, mechanics, and software.

Although robots vary in appearance and capabilities, they share the features of a mechanical, movable structure under some form of control. Generally, a robot's structure acts like the skeleton of a body and is called a "kinematic chain." The chain is formed of links (the robot's "bones"), actuators (the robot's "muscles"), and joints. In most contemporary robots, each link connects the one before it to the one after it. These structures often resemble the human arm. When a robot is needed to manipulate objects, the last link carries an "end effector," which can be anything from a mechanical hand to a welding device.

A robot's mechanical structure needs to be controlled on various levels, so that the device can recognize patterns, move along certain paths, avoid obstacles, and perform various tasks. The control of these functions usually involves three phases: perception, processing, and action (robotic paradigms). The robot has sensors that pick up information about the environment or the robot itself, such as the position of its joints or its end effector. Using strategies from the field of control theory, this information is processed to calculate the appropriate signals that need to be sent to the actuators (motors), and these in turn move the mechanical structure. Complex and adaptable control strategies may be referred to as artificial intelligence.

To perform any task, the robot needs the ability to move. When designing a robot, the roboticist needs to consider its motion in terms of (a) kinematics, or motion apart from the forces that cause the motion, and (b) dynamics, or motion in conjunction with the forces that affect it. The information gathered by studying the various possible motions can be used to improve the algorithms that control the robot's actions.

Robot

The term robot can be applied to any of a wide range of programmable machines, the common feature of which is that they are capable of movement and can be used to perform physical tasks. Robots are built in many different forms, ranging from humanoid to industrial, and their appearance is dictated by the functions they are to perform. They are usefully engaged in industrial production requiring repetitive tasks, and in performing tasks that may be difficult or dangerous for humans.

Several concerns have been expressed about robots. One is that they can displace people from their jobs. Another is that humanoid robots, if they acquire intelligence and emotions that allow them to think and act on their own, may become hostile toward humans and threaten humankind. The latter concern, however, is based on fictional concepts. Real-world robots are machines with their own mechanical limitations, and they lack the human spirit. Although they may be built to simulate human intelligence and decision-making traits to some extent, these abilities are governed by artificially programmed feedback loops, not by creative thought processes, heartfelt feelings, and senses of responsibility—qualities that underlie human aspirations and freedom of expression. Consequently, the abilities, activities, and value of robots will remain distinct from those of humans.

A humanoid robot manufactured by Toyota appears to be playing a trumpet.

A robot is a machine that can perform preprogrammed physical tasks. Robots have been built to serve various functions, and they therefore appear in a variety of forms. Some robotic devices, such as the robotic arm of the space shuttle, act under direct human control. Other robots act autonomously, under the control of a computer.

Certain robots, such as remotely controlled bomb-disposal units, are used to perform tasks that are too dangerous or difficult for humans to engage in directly. Others, including those used in automobile production, have made it possible to automate repetitive tasks, lowering the cost of performing them.

Some robots may act according to their own decision-making ability, provided by the technology of artificial intelligence. Such robots include feedback loops such that they can interact with and respond to their environment. They do not, however, display actual intelligence.

Types and uses of Robots

As robots grow in number and complexity, they are being more widely used in industry. Most often, they are used to perform repetitive tasks. Industrial robots can be manufactured in a wide range of sizes and can therefore handle much larger tasks than a human can. In addition, robots are useful in environments that are unpleasant or dangerous for people to work in, such as in toxic waste cleanup, bomb disposal, mining, and working in outer space or deep water.

Certain mobile robots, called automated guided vehicles (AGVs), are used in large facilities such as warehouses, hospitals, and container ports, for the transport of goods and for safety and security patrols. Such vehicles can be programmed to follow wires, markers, or laser guides to navigate around a given location.

KUKA Industrial robots assemble a vehicle underbody.

Some robots are available for performing domestic tasks such as cooking, vacuum cleaning, and lawn mowing. In addition, robot kits made of plastic or aluminum are becoming increasingly popular for education in schools and professional training by companies.

Humanoid robots are being developed with the aim of providing robotic functions in a form that may increase their appeal to customers. These types of robots, known as social robots, are made to interact with and provide companionship to people.

Industrial Robots

So far, the main use of robots has been in the automation of mass-production industries, where well-defined tasks must be repeated in exactly the same fashion, with little need for feedback to control the process. Typical applications of industrial robots include welding, painting, ironing, assembly, picking and placing, palletizing, product inspection, and testing. They can accomplish all these tasks with high endurance, speed, and precision. A prime example is automobile manufacture, in which large, complex robots are used.

The International Organization for Standardization (ISO) has officially defined an industrial robot as an "automatically controlled, reprogrammable, multipurpose manipulator, programmable in three or more axes." Most industrial robots would fall in the category of robotic arms, as implied by the term "manipulator" in this definition.

Industrial robots exhibit varying degrees of autonomy. Many are programmed to perform specific actions over and over again, without variation, and with a high degree of accuracy. Other industrial robots are built to be much more flexible about the orientation of the object on which they are operating or even the task that has to be performed, which the robot may need to identify. For example, these robots may contain machine vision subsystems acting as their "eyes," linked to powerful computers or controllers. Artificial intelligence, or what passes for it, is becoming an increasingly important factor in the modern industrial robot.

Humanoid Robots

A humanoid robot is a robot that is built to imitate the structure of the human body and to perform some of the same physical and mental tasks that humans can do. It is also built to function autonomously, in the sense of being able to adapt to changes in its environment or itself and continue to reach its goal. This is the main difference between humanoids and other types of robots, such as industrial robots, which perform tasks in highly structured environments. Consequently, humanoid robots are more complex than other kinds of robots and are more difficult to develop.

The capacities of a humanoid robot may include (a) autonomous learning, that is, learning new capabilities and adapting to new situations without outside assistance; (b) self-maintenance, such as the ability to recharge its batteries; (c) interacting safely with humans and the environment; and (d) avoiding situations that could harm others or itself.

To build and study humanoids, researchers need to understand the human body structure and behavior. Consequently, an important benefit of the work to develop humanoids is that it could lead to a better understanding of human biological and mental processes, from the seemingly simple act of walking to the concepts of consciousness and spirituality.

Research into humanoids is valuable in the medical field, such as in efforts to construct realistic leg and forearm prosthetic devices for amputees, or devices that help straighten deformed limbs. In addition, humanoid robots are being developed to assist the sick and elderly, or to perform dirty or dangerous jobs. The long-term goal of the creators of humanoid robots is that someday these devices will be able to understand human intelligence, reason and act like humans, and work alongside humans.

Humanoid robots are becoming increasingly popular in the realm of entertainment. For example, the female robot Ursula sings, dances, and speaks to her audiences at Universal Studios. Several Disney attractions employ the use of animatrons—robots that look, move, and speak like human beings—in some of their theme park shows. These animatrons look so realistic that it can be hard to decipher from a distance whether or not they are actually human. Despite their realistic appearance, they have no cognition or physical autonomy.

Machine

Wind turbines

The scientific definition of a machine is any device that transmits or modifies energy. In common usage, the meaning is restricted to devices having rigid moving parts that perform or assist in performing some work. Machines normally require some energy source ("input") and always accomplish some sort of work ("output"). Devices with no rigid moving parts are commonly considered tools, or simply devices, not machines.

People have used mechanisms to amplify their abilities since before written records were available. Generally these devices decrease the amount of force required to do a given amount of work, alter the direction of the force, or transform one form of motion or energy into another.

Modern power tools, automated machine tools, and human-operated power machinery are tools that are also machines. Machines used to transform heat or other energy into mechanical energy are known as engines.

Hydraulics devices may also be used to support industrial applications, although devices entirely lacking rigid moving parts are not commonly considered machines. Hydraulics are widely used in heavy equipment industries, automobile industries, marine industries, aeronautical industries, construction equipment industries, and earthmoving equipment industries.

Impact

Mechanization and Automation

A water-powered mine hoist used for raising ore. This woodblock is from De re metallica by Georg Bauer an early mining textbook that contains numerous drawings and descriptions of mining equipment.

Mechanization is providing human operators with machinery that assists them with the muscular requirements of work or displaces muscular work. In some fields, mechanization includes the use of hand tools. In modern usage, such as in engineering or economics, mechanization implies machinery more complex than hand tools and would not include simple devices such as an un-geared horse or donkey mill. Devices that cause speed changes or changes to or from reciprocating to rotary motion, using means such as gears, pulleys or sheaves and belts, shafts, cams and cranks, usually are considered machines. After electrification, when most small machinery was no longer hand powered, mechanization was synonymous with motorized machines.

Automation is the use of control systems and information technologies to reduce the need for human work in the production of goods and services. In the scope of industrialization, automation is a step beyond mechanization. Whereas mechanization provides human operators with machinery to assist them with the muscular requirements of work, automation greatly decreases the need for human sensory and mental requirements as well. Automation plays an increasingly important role in the world economy and in daily experience.

Automata

An automaton is a self-operating machine. The word is sometimes used to describe a robot, more specifically an autonomous robot.

Types

The mechanical advantage of a simple machine is the ratio between the force it exerts on the load and the input force applied. This does not entirely describe the machine's performance, as force is required to overcome friction as well. The mechanical efficiency of a machine is the ratio of the actual mechanical advantage (AMA) to the ideal mechanical advantage (IMA). Functioning physical machines are always less than 100 percent efficient.

Mechanical

The word mechanical refers to the work that has been produced by machines or the machinery. It mostly relates to the machinery tools and the mechanical applications of science. Some of its synonyms are automatic and mechanic.

Simple Machines

The idea that a machine can be broken down into simple movable elements led Archimedes to define the lever, pulley and screw as simple machines. By the time of the Renaissance this list increased to include the wheel and axle, wedge and inclined plane.

Engines

An engine or motor is a machine designed to convert energy into useful mechanical motion. Heat engines, including internal combustion engines and external combustion engines (such as steam engines) burn a fuel to create heat, which is then used to create motion. Electric motors convert electrical energy into mechanical motion, pneumatic motors use compressed air and others, such as wind-up toys use elastic energy. In biological systems, molecular motors like myosins in muscles use chemical energy to create motion.

Electrical

Electrical means operating by or producing electricity, relating to or concerned with electricity. In other words, it means using, providing, producing, transmitting or operated by electricity.

Electrical Machine

An electrical machine is the generic name for a device that converts mechanical energy to electrical energy, converts electrical energy to mechanical energy, or changes alternating current from one voltage level to a different voltage level.

Electronic Machine

Electronics is the branch of physics, engineering and technology dealing with electrical circuits that involve active electrical components such as vacuum tubes, transistors, diodes and integrated circuits, and associated passive interconnection technologies. The nonlinear behavior of active components and their ability to control electron flows makes amplification of weak signals possible and is usually applied to information and signal processing. Similarly, the ability of electronic devices to act as switches makes digital information processing possible. Interconnection technologies such as circuit boards, electronic packaging technology, and other varied forms of communication infrastructure complete circuit functionality and transform the mixed components into a working system.

Computing Machines

Computers are machines to process information, often in the form of numbers. Charles Babbage designed various machines to tabulate logarithms and other functions in 1837. His Difference engine can be considered an advanced mechanical calculator and his Analytical Engine a forerunner of the modern computer, though none were built in Babbage's lifetime.

Modern computers are electronic ones. They use electric charge, current or magnetization to store and manipulate information. Computer architecture deals with detailed design of computers. There are also simplified models of computers, like State machine and Turing machine.

Molecular Machines

Study of the molecules and proteins that are the basis of biological functions has led to the concept of a molecular machine. For example, current models of the operation of the kinesin molecule that transports vesicles inside the cell as well as the myosin molecule that operates against actin to cause muscle contraction; these molecules control movement in response to chemical stimuli.

Researchers in nano-technology are working to construct molecules that perform movement in response to a specific stimulus. In contrast to molecules such as kinesin and myosin, these nanomachines or molecular machines are constructions like traditional machines that are designed to perform in a task.

Types of machines and related components:

Classification	Machine(s)
Simple machines	Inclined plane, Wheel and axle, Lever, Pulley, Wedge, Screw
Mechanical components	Axle, Bearings, Belts, Bucket, Fastener, Gear, Key, Link chains, Rack and pinion, Roller chains, Rope, Seals, Spring, Wheel
Clock	Atomic clock, Watch, Pendulum clock, Quartz clock

<table>
<tr><td colspan="2">Compressors and Pumps</td><td>Archimedes' screw, Eductor-jet pump, Hydraulic ram, Pump, Trompe, Vacuum pump</td></tr>
<tr><td rowspan="2">Heat engines</td><td>External combustion engines</td><td>Steam engine, Stirling engine</td></tr>
<tr><td>Internal combustion engines</td><td>Reciprocating engine, Gas turbine</td></tr>
<tr><td colspan="2">Heat pumps</td><td>Absorption refrigerator, Thermoelectric refrigerator, Regenerative cooling</td></tr>
<tr><td colspan="2">Linkages</td><td>Pantograph, Cam, Peaucellier-Lipkin</td></tr>
<tr><td colspan="2">Turbine</td><td>Gas turbine, Jet engine, Steam turbine, Water turbine, Wind generator, Windmill</td></tr>
<tr><td colspan="2">Aerofoil</td><td>Sail, Wing, Rudder, Flap, Propeller</td></tr>
<tr><td colspan="2">Information technology</td><td>Computer, Calculator, Telecommunications networks</td></tr>
<tr><td colspan="2">Electricity</td><td>Vacuum tube, Transistor, Diode, Resistor, Capacitor, Inductor, Memristor, Semiconductor</td></tr>
<tr><td colspan="2">Robots</td><td>Actuator, Servo, Servomechanism, Stepper motor</td></tr>
<tr><td colspan="2">Miscellaneous</td><td>Vending machine, Wind tunnel, Check weighing machines, Riveting machines</td></tr>
</table>

Machine Elements

Machines are assembled from standardized types of components. These elements consist of mechanisms that control movement in various ways such as gear trains, transistor switches, belt or chain drives, linkages, cam and follower systems, brakes and clutches, and structural components such as frame members and fasteners.

Modern machines include sensors, actuators and computer controllers. The shape, texture and color of covers provide a styling and operational interface between the mechanical components of a machine and its users.

Mechanisms

Assemblies within a machine that control movement are often called "mechanisms." Mechanisms are generally classified as gears and gear trains, cam and follower mechanisms, and linkages, though there are other special mechanisms such as clamping linkages, indexing mechanisms and friction devices such as brakes and clutches.

Controllers

Controllers combine sensors, logic, and actuators to maintain the performance of components of a machine. Perhaps the best known is the flyball governor for a steam engine. Examples of these devices range from a thermostat that as temperature rises opens a valve to cooling water to speed controllers such the cruise control system in an automobile. The programmable logic controller replaced relays and specialized control mechanisms with a programmable computer. Servo motors that accurately position a shaft in response to an electrical command are the actuators that make robotic systems possible.

References

- Mechanical-Engineering, Engineering-branches: myklassroom.com, Retrieved 2 February, 2019
- Mechanical-engineering: britannica.com, Retrieved 17 May, 2019
- Mechatronics: techtarget.com, Retrieved 8 January, 2019
- Robot: newworldencyclopedia.org, Retrieved 3 August, 2019
- Machine: newworldencyclopedia.org, Retrieved 13 March, 2019

2

Mechanics: A Comprehensive Study

The branch of science which deals with the behavior of physical bodies when they are subjected to forces or displacements along with the effects which these bodies have on the environment is known as mechanics. This chapter discusses in detail the concepts related to mechanics such as motion, force and friction.

Mechanics is the science concerned with the motion of bodies under the action of forces, including the special case in which a body remains at rest. Of first concern in the problem of motion are the forces that bodies exert on one another. This leads to the study of such topics as gravitation, electricity, and magnetism, according to the nature of the forces involved. Given the forces, one can seek the manner in which bodies move under the action of forces; this is the subject matter of mechanics proper.

Historically, mechanics was among the first of the exact sciences to be developed. Its internal beauty as a mathematical discipline and its early remarkable success in accounting in quantitative detail for the motions of the Moon, the Earth, and other planetary bodies had enormous influence on philosophical thought and provided impetus for the systematic development of science into the 20th century.

Mechanics may be divided into three branches: statics, which deals with forces acting on and in a body at rest; kinematics, which describes the possible motions of a body or system of bodies; and kinetics, which attempts to explain or predict the motion that will occur in a given situation. Alternatively, mechanics may be divided according to the kind of system studied. The simplest mechanical system is the particle, defined as a body so small that its shape and internal structure are of no consequence in the given problem. More complicated is the motion of a system of two or more particles that exert forces on one another and possibly undergo forces exerted by bodies outside of the system.

The principles of mechanics have been applied to three general realms of phenomena. The motions of such celestial bodies as stars, planets, and satellites can be predicted with great accuracy thousands of years before they occur. (The theory of relativity predicts some deviations from the motion according to classical, or Newtonian, mechanics; however, these are so small as to be observable only with very accurate techniques, except in problems involving all or a large portion of the detectable universe.) As the second realm, ordinary objects on Earth down to microscopic size (moving at speeds much lower than that of light) are properly described by classical mechanics without significant corrections. The engineer who designs bridges or aircraft may use the Newtonian laws of classical mechanics with confidence, even though the forces may be very complicated, and the calculations lack the beautiful simplicity of celestial mechanics. The third realm of phenomena comprises the behaviour of matter and electromagnetic radiation on the atomic and subatomic scale. Although there were some limited early successes in describing the behaviour of atoms in terms of classical mechanics, these phenomena are properly treated in quantum mechanics.

Classical mechanics deals with the motion of bodies under the influence of forces or with the equilibrium of bodies when all forces are balanced. The subject may be thought of as the elaboration and application of basic postulates first enunciated by Isaac Newton They may be used to predict with great precision a wide variety of phenomena ranging from the motion of individual particles to the interactions of highly complex systems.

In the framework of modern physics, classical mechanics can be understood to be an approximation arising out of the more profound laws of quantum mechanics and the theory of relativity. However, that view of the subject's place greatly undervalues its importance in forming the context, language, and intuition of modern science and scientists. Our present-day view of the world and man's place in it is firmly rooted in classical mechanics. Moreover, many ideas and results of classical mechanics survive and play an important part in the new physics.

The central concepts in classical mechanics are force, mass, and motion. Neither force nor mass is very clearly defined by Newton, and both have been the subject of much philosophical speculation since Newton. Both of them are best known by their effects. Mass is a measure of the tendency of a body to resist changes in its state of motion. Forces, on the other hand, accelerate bodies, which is to say, they change the state of motion of bodies to which they are applied. The interplay of these effects is the principal theme of classical mechanics.

Although Newton's laws focus attention on force and mass, three other quantities take on special importance because their total amount never changes. These three quantities are energy, (linear) momentum, and angular momentum. Any one of these can be shifted from one body or system of bodies to another. In addition, energy may change form while associated with a single system, appearing as kinetic energy, the energy of motion; potential energy, the energy of position; heat, or internal energy, associated with the random motions of the atoms or molecules composing any real body; or any combination of the three. Nevertheless, the total energy, momentum, and angular momentum in the universe never changes. This fact is expressed in physics by saying that energy, momentum, and angular momentum are conserved. These three conservation laws arise out of Newton's laws, but Newton himself did not express them. They had to be discovered later.

It is a remarkable fact that, although Newton's laws are no longer considered to be fundamental, nor even exactly correct, the three conservation laws derived from Newton's laws—the conservation of energy, momentum, and angular momentum—remain exactly true even in quantum mechanics and relativity. In fact, in modern physics, force is no longer a central concept, and mass is only one of a number of attributes of matter. Energy, momentum, and angular momentum, however, still firmly hold centre stage. The continuing importance of these ideas inherited from classical mechanics may help to explain why this subject retains such great importance in science today.

Fundamental Concepts

Units and Dimensions

Quantities have both dimensions, which are an expression of their fundamental nature, and units, which are chosen by convention to express magnitude or size. For example, a series of events have a certain duration in time. Time is the dimension of the duration. The duration might be expressed as 30 minutes or as half an hour. Minutes and hours are among the units in which time may be expressed. One can compare quantities of the same dimensions, even if they are expressed in different units (an hour is longer than a minute). Quantities of different dimensions cannot be compared with one another.

The fundamental dimensions used in mechanics are time, mass, and length. Symbolically, these are written as t, m, and l, respectively. The study of electromagnetism adds an additional fundamental dimension, electric charge, or q. Other quantities have dimensions compounded of these. For example, speed has the dimensions distance divided by time, which can be written as l/t, and volume has the dimensions distance cubed, or l^3. Some quantities, such as temperature, have units but are not compounded of fundamental dimensions.

There are also important dimensionless numbers in nature, such as the number π = 3.14159. Dimensionless numbers may be constructed as ratios of quantities having the same dimension. Thus, the number π is the ratio of the circumference of a circle (a length) to its diameter (another length). Dimensionless numbers have the advantage that they are always the same, regardless of what set of units is being used.

Governments have traditionally been responsible for establishing and enforcing standard units for the sake of orderly commerce, navigation, science, and, of course, taxation. Today all such units are established by international treaty, revised every few years in light of scientific findings. The units used for most scientific measurements are those designated the International System of Units (Système International d'Unités) or SI for short. They are based on the metric system, first adopted officially by France in 1795. Other units, such as those of the British engineering system, are still in use in some places, but these are now defined in terms of the SI units.

The fundamental unit of length is the metre. A metre used to be defined as the distance between two scratch marks on a metal bar kept in Paris, but it is now much more precisely defined as the distance that light travels in a certain time interval (1/299,792,458 of a second). By contrast, in the British system, units of length have a clear human bias: the foot, the inch (the first joint of the thumb) the yard (distance from nose to outstretched fingertip), and the mile (one thousand standard paces of a Roman legion). Each of these is today defined as some fraction or multiple of a metre (one yard is nearly equal to one metre). In the SI or the metric system, lengths are expressed as decimal fractions or multiples of a metre (a millimetre = one-thousandth of a metre; a centimetre = one-hundredth of a metre; a kilometre = one thousand metres).

Times longer than one second are expressed in the units seconds, minutes, hours, days, weeks, and years. Times shorter than one second are expressed as decimal fractions (a millisecond = one-thousandth of a second, a microsecond = one-millionth of a second, and so on). The fundamental unit of time is today based on the intrinsic properties of certain kinds of atoms (an excitation frequency of the isotope cesium-133).

Units of mass are also defined in a way that is technically sound, but in common usage they are the subject of some confusion because they are easily confused with units of weight, which is a different physical quantity. The weight of an object is the consequence of the Earth's gravity operating on its mass. Thus, the mass of a given object is the same everywhere, but its weight varies slightly if it is moved about the surface of the Earth, and it would change a great deal if it were moved to the surface of another planet. Also, weight and mass do not have the same dimensions (weight has the dimensions ml/t^2). The Constitution of the United States, which calls on the government to establish uniform "weights and measures," is oblivious to this distinction, as are merchants the world over, who measure the weight of bread or produce but sell it in units of kilograms, the SI unit of mass. (The kilogram is equal to 1,000 grams; 1 gram is the mass of 1 cubic centimetre of water—under appropriate conditions of temperature and pressure).

Vectors

The equations of mechanics are typically written in terms of Cartesian coordinates. At a certain time t, the position of a particle may be specified by giving its coordinates x(t), y(t), and z(t) in a particular Cartesian frame of reference. However, a different observer of the same particle might choose a differently oriented set of mutually perpendicular axes, say, x′, y′, and z′. The motion of the particle is then described by the first observer in terms of the rate of change of x(t), y(t), and z(t), while the second observer would discuss the rates of change of x′(t), y′(t), and z′(t). That is, both observers see the same particle executing the same motion and obeying the same laws, but they describe the situation with different equations. This awkward situation may be avoided by means of a mathematical construction called a vector. Although vectors are mathematically simple and extremely useful in discussing mechanics, they were not developed in their modern form until late in the 19th century, when J. Willard Gibbs and Oliver Heaviside (of the United States and Britain, respectively) each applied vector analysis in order to help express the new laws of electromagnetism proposed by James Clerk Maxwell.

A vector is a quantity that has both magnitude and direction. It is typically represented symbolically by an arrow in the proper direction, whose length is proportional to the magnitude of the vector. Although a vector has magnitude and direction, it does not have position. A vector is not altered if it is displaced parallel to itself as long as its length is not changed.

By contrast to a vector, an ordinary quantity having magnitude but not direction is known as a scalar. In printed works vectors are often represented by boldface letters such as A or X, and scalars are represented by lightface letters, A or X. The magnitude of a vector, denoted|A|, is itself a scalar—i.e.,|A|= A.

Because vectors are different from ordinary (i.e., scalar) quantities, all mathematical operations involving vectors must be carefully defined. Addition, subtraction, three kinds of multiplication, and differentiation will be discussed here. There is no mathematical operation that corresponds to division by a vector.

If vector A is added to vector B, the result is another vector, C, written A + B = C. The operation is performed by displacing B so that it begins where A ends, as shown in Figure. C is then the vector that starts where A begins and ends where B ends.

Vector addition is defined to have the (nontrivial) property A + B = B + A. There do exists quantities having magnitude and direction that do not obey this requirement. An example is finite rotations in space. Two finite rotations of a body about different axes do not necessarily result in the same orientation if performed in the opposite order.

Vector subtraction is defined by A – B = A + (–B), where the vector –B has the same magnitude as B but the opposition direction. The idea is illustrated in the figure.

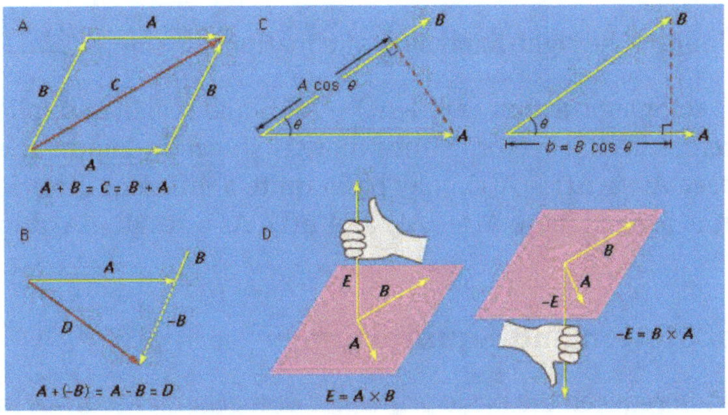

Figure: (A) The vector sum C = A + B = B + A. (B) The vector difference A + (−B) = A − B = D. (C, left) A cos θ is the component of A along B and (right) B cos θ is the component of B along A. (D, left) The right-hand rule used to find the direction of E = A × B and (right) the right-hand rule used to find the direction of −E = B × A.

A vector may be multiplied by a scalar. Thus, for example, the vector 2A has the same direction as A but is twice as long. If the scalar has dimensions, the resulting vector still has the same direction as the original one, but the two cannot be compared in magnitude. For example, a particle moving with constant velocity v suffers a displacement s in time t given by s = vt. The vector v has been multiplied by the scalar t to give a new vector, s, which has the same direction as v but cannot be compared to v in magnitude (a displacement of one metre is neither bigger nor smaller than a velocity of one metre per second). This is a typical example of a phenomenon that might be represented by different equations in differently oriented Cartesian coordinate systems but that has a single vector equation (for all observers not moving with respect to one another).

The dot product (also known as the scalar product, or sometimes the inner product) is an operation that combines two vectors to form a scalar. The operation is written A · B. If θ is the (smaller) angle between A and B, then the result of the operation is A · B = AB cos θ. The dot product measures the extent to which two vectors are parallel. It may be thought of as multiplying the magnitude of one vector (either one) by the projection of the other upon it, as shown in Figure. If the two vectors are perpendicular, the dot product is zero.

The cross product (also known as the vector product) combines two vectors to form another vector, perpendicular to the plane of the original vectors. The operation is written A × B. If θ is the (smaller) angle between A and B, then $|A \times B| = AB \sin\theta$. The direction of A × B is given by the right-hand rule: if the fingers of the right hand are made to rotate from A through θ to B, the thumb points in the direction of A × B, as shown in figure. The cross product is zero if the two vectors are parallel, and it is maximum in magnitude if they are perpendicular.

The derivative or rate of change, of a vector is defined in perfect analogy to the derivative of a scalar: if the vector A changes with time t, then

$$\frac{dA}{dt} = \lim_{\Delta t \to 0} \frac{A(t + \Delta t) - A(t)}{\Delta t}.$$

Before going to the limit on the right-hand side of equation $\frac{dA}{dt} = \lim_{\Delta t \to 0} \frac{A(t+\Delta t) - A(t)}{\Delta t}$, the operations described are vector subtraction [A(t + Δt) – A(t)] and scalar multiplication (by 1/Δt). The result, dA/dt, is therefore itself a vector. Notice that, as shown in Figure, the difference between two vectors, in this case A(t + Δt) – A(t), may be in quite a different direction than either of the vectors from which it is formed, here A(t + Δt) and A(t). As a result, dA/dt may be in a different direction than A(t).

Newton's Laws of Motion and Equilibrium

In his *Principia*, Newton reduced the basic principles of mechanics to three laws:

1. Everybody continues in its state of rest or of uniform motion in a straight line, unless it is compelled to change that state by forces impressed upon it.
2. The change of motion of an object is proportional to the force impressed and is made in the direction of the straight line in which the force is impressed.
3. To every action there is always opposed an equal reaction; or, the mutual actions of two bodies upon each other are always equal and directed to contrary parts.

Newton's first law is a restatement of the principle of inertia, proposed earlier by Galileo and perfected by Descartes.

The second law is the most important of the three; it may be understood very nearly to summarize all of classical mechanics. Newton used the word "motion" to mean what is today called momentum—that is, the product of mass and velocity, or p = mv, where p is the momentum, m the mass, and v the velocity of a body. The second law may then be written in the form of the equation F = dp/dt, where F is the force, the time derivative expresses Newton's "change of motion," and the vector form of the equation assures that the change is in the same direction as the force, as the second law requires.

For a body whose mass does not change,

$$\frac{dp}{dt} = m\frac{dv}{dt} = ma.$$

where a is the acceleration. Thus, Newton's second law may be put in the following form:

$$F = ma.$$

It is probably fair to say that equation $F = ma$ is the most famous equation in all of physics.

Newton's third law assures that when two bodies interact, regardless of the nature of the interaction, they do not produce a net force acting on the two-body system as a whole. Instead, there is an action and reaction pair of equal and opposite forces, each acting on a different body (action and reaction forces never act on the same body). The third law applies whether the bodies in question are at rest, in uniform motion, or in accelerated motion.

If a body has a net force acting on it, it undergoes accelerated motion in accordance with the second law. If there is no net force acting on a body, either because there are no forces at all or because all forces are precisely balanced by contrary forces, the body does not accelerate and may be said to be in equilibrium. Conversely, a body that is observed not to be accelerated may be deduced to have no net force acting on it.

Consider, for example, a massive object resting on a table. The object is known to be acted on by the gravitational force of the Earth; if the table were removed, the object would fall. It follows therefore from the fact that the object does not fall that the table exerts an upward force on the object, equal and opposite to the downward force of gravity. This upward force is not a mere physicist's bookkeeping device but rather a real physical force. The table's surface is slightly deformed by the weight of the object, causing the surface to exert a force analogous to that exerted by a coiled spring.

It is useful to recall the following distinction. The massive object exerts a downward force on the table that is equal and opposite to the upward force exerted by the table (owing to its deformation) on the object. These two forces are an action and reaction pair operating on different bodies (one on the table, the other on the object) as required by Newton's third law. On the other hand, the upward force exerted on the object by the table is balanced by a downward force exerted on the object by the Earth's gravity. These two equal and opposite forces, acting on the same body, are not related to or by Newton's third law, but they do produce the equilibrium immobile state of the body.

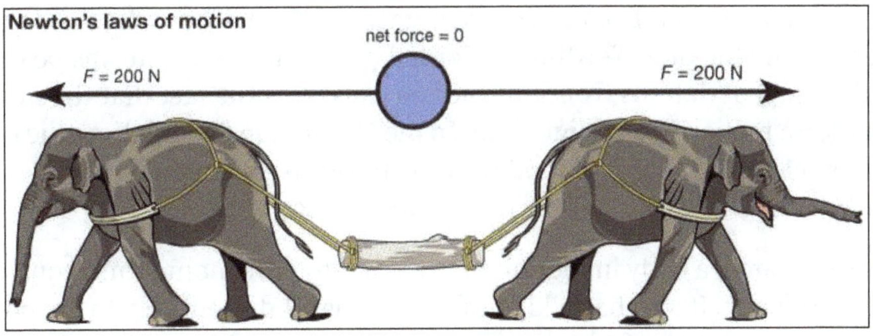

When an object is acted on by equal and opposite forces, it is at rest.

Motion

The continuous change in position of a body with respect to time and relative to the reference point or observer is called motion. Here the reference point can be considered as stationary or motionless objects surrounding the body.

Example: Consider sun as reference point or observer the planets orbiting the sun are said to be in motion.

The position of a body is not changing with time and relative to the reference point then the body is said to be at rest or motionless. Chairs, tables, Television are some examples of objects which are at rest. Rest and motion are relative terms for example consider a person in a moving bus he is at rest with respect to another person in the moving bus, but he is in motion with respect person outside the bus.

Motion of a Particle in One Dimension

Uniform Motion

According to Newton's first law (also known as the principle of inertia), a body with no net force acting on it will either remain at rest or continue to move with uniform speed in a straight line, according to its initial condition of motion. In fact, in classical Newtonian mechanics, there is no important distinction between rest and uniform motion in a straight line; they may be regarded as the same state of motion seen by different observers, one moving at the same velocity as the particle, the other moving at constant velocity with respect to the particle.

Although the principle of inertia is the starting point and the fundamental assumption of classical mechanics, it is less than intuitively obvious to the untrained eye. In Aristotelian mechanics, and in ordinary experience, objects that are not being pushed tend to come to rest. The law of inertia was deduced by Galileo from his experiments with balls rolling down inclined planes, described above.

For Galileo, the principle of inertia was fundamental to his central scientific task. He had to explain how it is possible that if Earth is really spinning on its axis and orbiting the Sun we do not sense that motion. The principle of inertia helps to provide the answer – since we are in motion together with Earth, and our natural tendency is to retain that motion, Earth appears to us to be at rest. Thus, the principle of inertia, far from being a statement of the obvious, was once a central issue of scientific contention. By the time Newton had sorted out all the details, it was possible to account accurately for the small deviations from this picture caused by the fact that the motion of Earth's surface is not uniform motion in a straight line. In the Newtonian formulation, the common observation that bodies that are not pushed tend to come to rest is attributed to the fact that they have unbalanced forces acting on them, such as friction and air resistance.

As has already been stated, a body in motion may be said to have momentum equal to the product of its mass and its velocity. It also has a kind of energy that is due entirely to its motion, called kinetic energy. The kinetic energy of a body of mass m in motion with velocity v is given by,

$$K = \frac{1}{2} m v^2 .$$

Falling Bodies and Uniformly Accelerated Motion

During the 14th century, the French scholar Nicole Oresme studied the mathematical properties of uniformly accelerated motion. He had little interest in whether that kind of motion could be observed in the realm of actual human existence, but he did discover that, if a particle is uniformly accelerated, its speed increases in direct proportion to time, and the distance it traverses is proportional to the square of the time spent accelerating. Two centuries later, Galileo repeated these same mathematical discoveries (perhaps independently) and, just as important, determined that this kind of motion is actually executed by balls rolling down inclined planes. As the incline of the plane increases, the acceleration increases, but the motion continues to be uniformly accelerated.

From this observation, Galileo deduced that a body falling freely in the vertical direction would also have uniform acceleration. Even more remarkably, he demonstrated that, in the absence of air resistance, all bodies would fall with the same constant acceleration regardless of their mass. If the constant acceleration of anybody dropped near the surface of Earth is expressed as g, the behaviour of a body dropped from rest at height zo and time t = 0 may be summarized by the following equations:

$$z = z_0 - \frac{1}{2}gt^2$$

$$v = gt,$$

$$a = g,$$

where z is the height of the body above the surface, v is its speed, and a is its acceleration. These equations of motion hold true until the body actually strikes the surface. The value of g is approximately 9.8 metres per second squared (m/s^2).

A body of mass m at a height z_0 above the surface may be said to possess a kind of energy purely by virtue of its position. This kind of energy (energy of position) is called potential energy. The gravitational potential energy is given by,

$$U = mgz_0.$$

Technically, it is more correct to say that this potential energy is a property of the Earth-body system rather than a property of the body itself, but this pedantic distinction can be ignored.

As the body falls to height z less than z_0, its potential energy U converts to kinetic energy $K = {}^1/_2 mv^2$. Thus, the speed v of the body at any height z is given by solving the equation,

$$\frac{1}{2}mv^2 + mgz = mgz_0.$$

Equation $\frac{1}{2}mv^2 + mgz = mgz_0$ is an expression of the law of conservation of energy. It says that the sum of kinetic energy, ${}^1/_2 mv^2$, and potential energy, mgz, at any point during the fall, is equal to the total initial energy, mgz_0, before the fall began. Exactly the same dependence of speed on height could be deduced from the kinematic equations $z = z_0 - \frac{1}{2}gt^2$, $v = gt$ and $a = g$, above.

In order to reach the initial height z_0, the body had to be given its initial potential energy by some external agency, such as a person lifting it. The process by which a body or a system obtains mechanical energy from outside of itself is called work. The increase of the energy of the body is equal to the work done on it. Work is equal to force times distance.

The force exerted by Earth's gravity on a body of mass m may be deduced from the observation that the body, if released, will fall with acceleration g. Since force is equal to mass times acceleration, the force of gravity is given by F = mg. To lift the body to height z_0, an equal and opposite (i.e., upward) force must be exerted through a distance z_0. Thus, the work done is,

$$W = Fz_0 = mgz_0,$$

which is equal to the potential energy that results.

If work is done by applying a force to a body that is not being acted upon by an opposing force, the body is accelerated. In this case, the work endows the body with kinetic energy rather than potential energy. The energy that the body gains is equal to the work done on it in either case. It should be noted that work, potential energy, and kinetic energy, all being aspects of the same quantity, must all have the dimensions ml^2/t^2.

Simple Harmonic Oscillations

Consider a mass m held in an equilibrium position by springs, as shown in Figure. The mass may be perturbed by displacing it to the right or left. If x is the displacement of the mass from equilibrium, the springs exert a force F proportional to x, such that where k is a constant that depends on the stiffness of the springs.

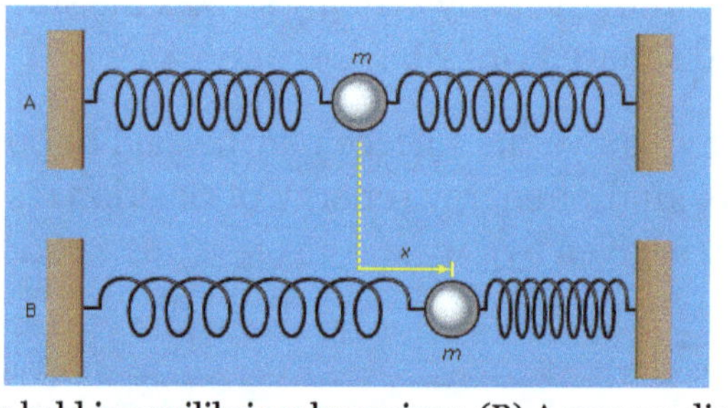

Figure: (A) A mass m held in equilibrium by springs. (B) A mass m displaced a distance x.

Equation $F = -kx$ is called Hooke's law, and the force is called the spring force. If x is positive (displacement to the right), the resulting force is negative (to the left), and vice versa. In other words, the spring force always acts so as to restore mass back toward its equilibrium position. Moreover, the force will produce an acceleration along the x direction given by $a = d^2x/dt^2$. Thus, Newton's second law, F = ma, is applied to this case by substituting $-kx$ for F and d^2x/dt^2 for a, giving $-kx = m(d^2x/dt^2)$. Transposing and dividing by m yields the equation:

$$F = -kx.$$

Equation $a = \frac{d^2x}{dt^2} = -\frac{k}{m}x$ gives the derivative—in this case the second derivative—of a quantity x in terms of the quantity itself. Such an equation is called a differential equation, meaning an equation containing derivatives. Much of the ordinary, day-to-day work of theoretical physics consists of solving differential equations. The question is, given equation $a = \frac{d^2x}{dt^2} = -\frac{k}{m}x$, how does x depend on time?

The answer is suggested by experience. If the mass is displaced and released, it will oscillate back and forth about its equilibrium position. That is, x should be an oscillating function of t, such as a sine wave or a cosine wave. For example, x might obey a behaviour such as:

$$x = A\cos\omega t.$$

Equation $x = A\cos\omega t$ describes the behaviour sketched graphically in Figure. The mass is initially displaced a distance x = A and released at time t = 0. As time goes on, the mass oscillates from A to −A and back to A again in the time it takes ωt to advance by 2π. This time is called T, the period of oscillation, so that ωT = 2π, or T = 2π/ω. The reciprocal of the period, or the frequency f, in oscillations per second, is given by f = 1/T = ω/2π. The quantity ω is called the angular frequency and is expressed in radians per second:

$$x = A\cos\,\omega t.$$

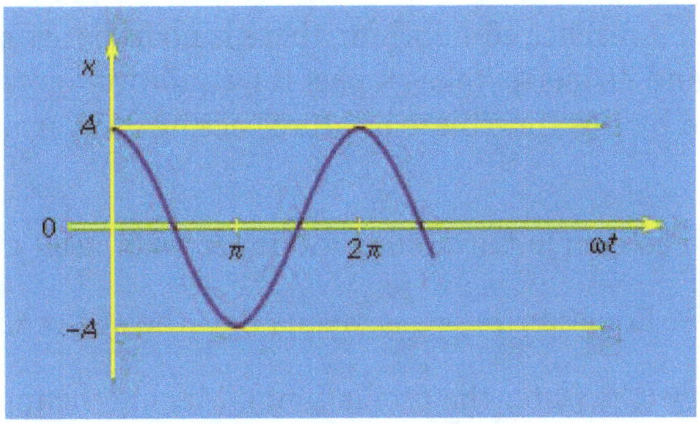

The function x = A cos ωt.

The choice of equation $x = A\cos\,\omega t$ as a possible kind of behaviour satisfying the differential equation $a = \dfrac{d^2x}{dt^2} = -\dfrac{k}{m}x$ can be tested by substituting it into equation $a = \dfrac{d^2x}{dt^2} = -\dfrac{k}{m}x$. The first derivative of x with respect to t is,

$$\frac{dx}{dt} = \frac{d}{dt}(A\cos\,\omega t)$$
$$= -\omega A\sin\omega t$$

Differentiating a second time gives,

$$\frac{d^2x}{dt^2} = \frac{d}{dt}\left(\frac{dx}{dt}\right)$$
$$= \frac{d}{dt}(-\omega A\cos\omega t)$$
$$= -\omega^2 A\cos\omega t$$
$$= -\omega^2 x.$$

Equation $\dfrac{d^2x}{dt^2} = \dfrac{d}{dt}\left(\dfrac{dx}{dt}\right)$ is the same as equation $a = \dfrac{d^2x}{dt^2} = -\dfrac{k}{m}x$. If,

$$\omega^2 = \frac{k}{m}.$$

Thus, subject to this condition, equation $x = A\cos\,\omega t$ is a correct solution to the differential equation. There are other possible correct guesses (e.g., x = A sin ωt) that differ from this one only in whether the mass is at rest or in motion at the instant t = 0.

The mass, as has been shown, oscillates from A to –A and back again. The speed, given by dx/dt, equation $\frac{dx}{dt} = \frac{d}{dt}\left(A\cos\omega t\right) = -\omega A\sin\omega t$ is zero at A and –A, but has its maximum magnitude, equal to ωA, when x is equal to zero. Physically, after the mass is displaced from equilibrium a distance A to the right, the restoring force F pushes the mass back toward its equilibrium position, causing it to accelerate to the left. When it reaches equilibrium, there is no force acting on it at that instant, but it is moving at speed ωA, and its inertia takes it past the equilibrium position. Before it is stopped it reaches position –A, and by this time there is a force acting on it again, pushing it back toward equilibrium.

The whole process, known as simple harmonic motion, repeats itself endlessly with a frequency given by equation $\omega^2 = \frac{k}{m}$. Equation $\omega^2 = \frac{k}{m}$. means that the stiffer the springs (i.e., the larger k), the higher the frequency (the faster the oscillations). Making the mass greater has exactly the opposite effect, slowing things down.

One of the most important features of harmonic motion is the fact that the frequency of the motion, ω (or f), depends only on the mass and the stiffness of the spring. It does not depend on the amplitude A of the motion. If the amplitude is increased, the mass moves faster, but the time required for a complete round trip remains the same. This fact has profound consequences, governing the nature of music and the principle of accurate timekeeping.

The potential energy of a harmonic oscillator, equal to the work an outside agent must do to push the mass from zero to x, is $U = \frac{1}{2}kx^2$. Thus, the total initial energy in the situation described above is $\frac{1}{2}kA^2$; and since the kinetic energy is always $\frac{1}{2}mv^2$, when the mass is at any point x in the oscillation,

$$\frac{1}{2}mv^2 + \frac{1}{2}kx^2 = \frac{1}{2} = kA^2.$$

Equation $\frac{1}{2}mv^2 + \frac{1}{2}kx^2 = \frac{1}{2} = kA^2$ plays exactly the role for harmonic oscillators that equation $\frac{1}{2}mv^2 + mgz = mgz_0$ does for falling bodies.

It is quite generally true that harmonic oscillations result from disturbing anybody or structure from a state of stable mechanical equilibrium.

Consider a bowl with a marble resting inside, then consider a second, inverted bowl with a marble balanced on top. In both cases, the net force on the marble is zero. The marbles are thus in mechanical equilibrium. However, a small disturbance in the position of the marble balanced on top of the inverted bowl will cause it to roll away and not return. In such a case, the equilibrium is said to be unstable. Conversely, if the marble inside the first bowl is disturbed, gravity acts to push it back toward the bottom of the bowl. The marble inside the bowl is an example of a body in stable equilibrium. If it is disturbed slightly, it executes harmonic oscillations around the bottom of the bowl rather than rolling away.

This argument may be generalized by a simple mathematical argument. Consider a body or structure in mechanical equilibrium, which, when disturbed by a small amount x, finds a force acting on it that is a function of x, F(x). For small x, such a function may be written generally as a power series in x; i.e.,

$$F(x) = F(0) + ax + bx^2 + ...,$$

where F(0) is the value of F(x) when x = (0), and a and b are constants, independent of x, determined by the nature of the system. The statement that the body is in mechanical equilibrium means that F(0) = 0, so that no force is acting on the body when it is undisturbed. Since x is small, x^2 is much smaller; thus, the term bx^2 and all higher powers may be disregarded. This leaves F(x) = ax. Now, if a is positive, a disturbance produces a force in the same direction as the disturbance. This was the case when the marble was balanced on top of the inverted bowl. It describes unstable equilibrium. For the system to be stable, a must be negative. Thus, if a = −k, where k is some positive constant, equation $F(x) = F(0) + ax + bx^2 + ...,$ becomes F(x) = −kx, which is simply Hooke's law, equation $F = -kx,$. As has been described above, any system obeying Hooke's law is a harmonic oscillator.

The generality of this argument accounts for the fact that harmonic oscillators are abundantly observed in common experience. For example, any rigid structure will oscillate at many different harmonic frequencies corresponding to different possible distortions of its equilibrium shape. In addition, music may be produced either by disturbing the equilibrium of a stretched wire or fibre (as in the piano and violin), a stretched membrane (e.g., drums), or a rigid bar (the triangle and the xylophone) or by disturbing the density of an enclosed column of air (as in the trumpet and organ). While a fluid such as air is not rigid, its density is an example of a stable system that obeys Hooke's law and may therefore be set into harmonic oscillations.

All music would be quite different from what it is was it not for the general property of harmonic oscillators that the frequency is independent of the amplitude. Thus, instruments yield the same note (frequency) regardless of how loudly they are played (amplitude), and, equally important, the same note persists as the vibrations die away. This same property of harmonic oscillators is the underlying principle of all accurate timekeeping.

The first precise timekeeping mechanism, whose principles of motion were discovered by Galileo, was the simple pendulum. The accuracy of modern timekeeping has been improved dramatically by the introduction of tiny quartz crystals, whose harmonic oscillations generate electrical signals that may be incorporated into miniaturized circuits in clocks and wristwatches. All harmonic oscillators are natural timekeeping devices because they oscillate at intrinsic natural frequencies independent of amplitude. A given number of complete cycles always corresponds to the same elapsed time. Quartz crystal oscillators make more accurate clocks than pendulums do principally because they oscillate many more times per second.

Damped and Forced Oscillations

The simple harmonic oscillations continue forever, at constant amplitude, oscillating as shown in Figure between A and −A. Common experience indicates that real oscillators behave somewhat differently, however. Harmonic oscillations tend to die away as time goes on. This behaviour, called damping of the oscillations, is produced by forces such as friction and viscosity. These forces are known collectively as dissipative forces because they tend to dissipate the potential and kinetic energies of macroscopic bodies into the energy of the chaotic motion of atoms and molecules known as heat.

Friction and viscosity are complicated phenomena whose effects cannot be represented accurately by a general equation. However, for slowly moving bodies, the dissipative forces may be represented by,

$$F_d = -\gamma v,$$

where v is the speed of the body and γ is a constant coefficient, independent of dynamic quantities such as speed or displacement. Equation $F_d = -\gamma v$, is most easily understood by an argument analogous to that applied to equation $F(x) = F(0) + ax + bx^2 + ...$, above. F_d is written as a sum of powers of v, or $F_d(v) = F_d(0) + av + bv^2 + \cdots$. When the body is at rest (v = 0), no dissipative force is expected because, if there were one, it might set the body into motion. Thus, $F_d(0) = 0$. The next term must be negative since dissipative forces always resist the motion. Thus, a = −γ where γ is positive. Since v^2 has the same sign regardless of the direction of the motion, b must equal 0 lest it sometimes contribute a dissipative force in the same direction as the motion. The next term is proportional to v^3, and it and all subsequent terms may be neglected if v is sufficiently small. So, as in equation $F(x) = F(0) + ax + bx^2 + ...$ the power series is reduced to a single term, in this case $F_d = -\gamma v$.

To find the effect of a dissipative force on a harmonic oscillator, a new differential equation must be solved. The net force, or mass times acceleration, written as md^2x/dt^2, is set equal to the sum of the Hooke's law force, −kx, and the dissipative force, −γv = −γdx/dt. Dividing by m yields,

$$\frac{d^2x}{dt^2} = -\frac{k}{m}x - \frac{\gamma}{m}\frac{dx}{dt}.$$

The general solution to equation $\frac{d^2x}{dt^2} = -\frac{k}{m}x - \frac{\gamma}{m}\frac{dx}{dt}$ is given in the form $x = Ce^{-\gamma t/2m}\cos(\omega t + \theta_0)$,

where C and θ_0 are arbitrary constants determined by the initial conditions. This motion, for the case in which $\theta_0 = 0$, is illustrated in Figure. As expected, the harmonic oscillations die out with time. The amplitude of the oscillations is bounded by an exponentially decreasing function of time (the dashed curves). The characteristic decay time (after which the oscillations are smaller by 1/e, where e is the base of the natural logarithms e = 2.718 . . .) is equal to 2m/γ. The frequency of the oscillations is given in figure.

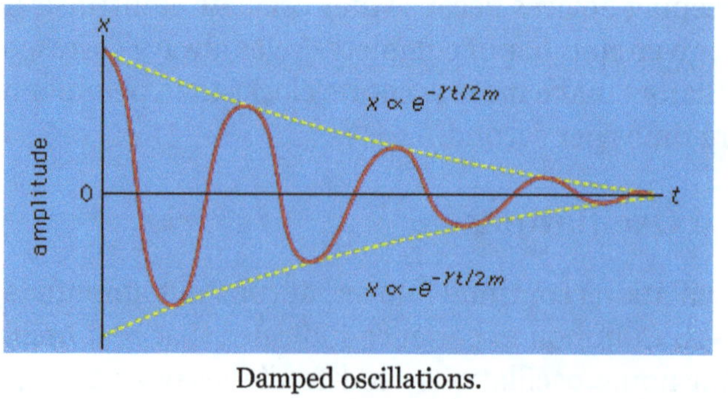

Damped oscillations.

Importantly, this frequency does not change as the oscillations decay.

Equation $\omega^2 = \frac{k}{m} - \frac{\gamma^2}{4m^2}$ shows that it is possible, by proper choice of γ, to turn a harmonic oscillator into a system that does not oscillate at all—that is, a system whose natural frequency is ω = 0. Such a system is said to be critically damped. For example, the springs that suspend the body of an automobile cause it to be a natural harmonic oscillator. The shock absorbers of the auto are devices that seek to add just enough dissipative force to make the assembly critically damped. In this way, the passengers need not go through numerous oscillations after each bump in the road.

A simple disturbance can set a harmonic oscillator into motion. Repeated disturbances can increase the amplitude of the oscillations if they are applied in synchrony with the natural frequency. Even a very small disturbance, repeated periodically at just the right frequency, can cause a very large amplitude motion to build up. This phenomenon is known as resonance.

Periodically forced oscillations may be represented mathematically by adding a term of the form a_0 sin ωt to the right-hand side of equation $\omega^2 = \frac{k}{m} - \frac{\gamma^2}{4m^2}$.

This term describes a force applied at frequency ω, with amplitude ma_0. The result of applying such a force is to create a kind of motion that does not need to decay with time, since the energy lost to dissipative processes is replaced, over the course of each cycle, by the driving force. The amplitude of the motion depends on how close the driving frequency ω is to the natural frequency ω_0 of the oscillator. Interestingly, even though dissipation is present, ω_0 is not given by equation $\omega^2 = \frac{k}{m} - \frac{\gamma^2}{4m^2}$ but rather by equation $\omega^2 = \frac{k}{m}$, $\omega^2_0 = k/m$.

In a graph of the amplitude of the steady state motion (i.e., long after the driving force has begun to be applied), the maximum amplitude occurs as expected at $\omega = \omega_0$. The height and width of the resonance curve are governed by the damping coefficient γ. If there were no damping, the maximum amplitude would be infinite. Because small disturbances at every possible frequency are always present in the natural world, every rigid structure would shake itself to pieces if not for the presence of internal damping.

Resonances are not uncommon in the world of familiar experience. For example, cars often rattle at certain engine speeds, and windows sometimes rattle when an airplane flies by. Resonance is particularly important in music. For example, the sound box of a violin does its job well if it has a natural frequency of oscillation that responds resonantly to each musical note. Very strong resonances to certain notes—called "wolf notes" by musicians—occur in cheap violins and are much to be avoided. Sometimes, a glass may be broken by a singer as a result of its resonant response to a particular musical note.

Motion of a Particle in Two or more Dimensions

Projectile Motion

Galileo was quoted above pointing out with some detectable pride that none before him had realized that the curved path followed by a missile or projectile is a parabola. He had arrived at his conclusion by realizing that a body undergoing ballistic motion executes, quite independently, the motion of a freely falling body in the vertical direction and inertial motion in the horizontal direction. These considerations, and terms such as ballistic and projectile, apply to a body that, once launched, is acted upon by no force other than Earth's gravity.

Projectile motion may be thought of as an example of motion in space that is to say, of three-dimensional motion rather than motion along a line, or one-dimensional motion. In a suitably defined system of Cartesian coordinates, the position of the projectile at any instant may be specified by giving the values of its three coordinates, x(t), y(t), and z(t). By generally accepted convention, z(t) is used to describe the vertical direction. To a very good approximation, the motion is confined to a single vertical plane, so that for any single projectile it is possible to choose a coordinate system such that the motion is two-dimensional [say, x(t) and z(t)] rather than three-dimensional [x(t), y(t), and z(t)]. It is assumed throughout this section that the range of the motion is sufficiently limited that the curvature of Earth's surface may be ignored.

Consider a body whose vertical motion obeys equation $z = z_0 - \frac{1}{2}gt^2$, Galileo's law of falling bodies, which states $z = z_0 - {}^1/_2 gt^2$, while, at the same time, moving horizontally at a constant speed v_x in accordance with Galileo's law of inertia. The body's horizontal motion is thus described by $x(t) = v_x t$, which may be written in the form $t = x/v_x$.

Using this result to eliminate t from equation $z = z_0 - \frac{1}{2}gt^2$ gives $z = z_0 - {}^1/_2 g(1/vx)^2x^2$. This latter is the equation of the trajectory of a projectile in the z–x plane, fired horizontally from an initial height z_0. It has the general form where a and b are constants. Equation $z = a + bx^2$ may be recognized to describe a parabola just as Galileo claimed. The parabolic shape of the trajectory is preserved even if the motion has an initial component of velocity in the vertical direction.

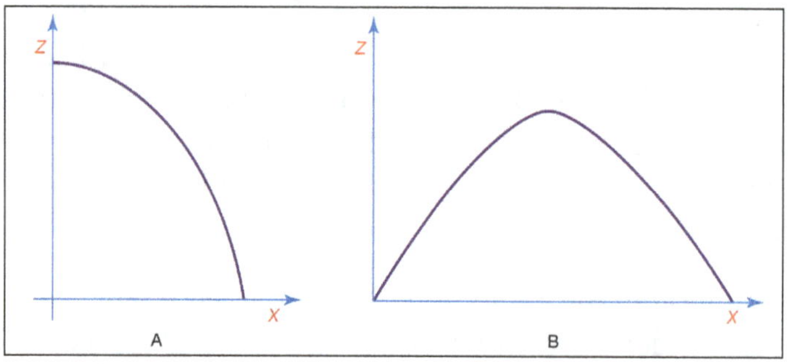

(A) The parabolic path of a projectile. (B) The parabolic path of a projectile with an initial upward component of velocity.

Energy is conserved in projectile motion. The potential energy U(z) of the projectile is given by U(z) = mgz. The kinetic energy K is given by $K = {}^1/_2 mv^2$, where v^2 is equal to the sum of the squares of the vertical and horizontal components of velocity, or $v^2 = v^2x + v^2z$.

In all of this discussion, the effects of air resistance (to say nothing of wind and other more complicated phenomena) have been neglected. These effects are seldom actually negligible. They are most nearly so for bodies that are heavy and slow-moving. All of this discussion, therefore, is of great value for understanding the underlying principles of projectile motion but of little utility for predicting the actual trajectory of, say, a cannonball once fired or even a well-hit baseball.

Motion of a Pendulum

According to legend, Galileo discovered the principle of the pendulum while attending mass at the Duomo (cathedral) located in the Piazza del Duomo of Pisa, Italy. A lamp hung from the ceiling by a cable and, having just been lit, was swaying back and forth. Galileo realized that each complete cycle of the lamp took the same amount of time, compared to his own pulse, even though the amplitude of each swing was smaller than the last. As has already been shown, this property is common to all harmonic oscillators, and, indeed, Galileo's discovery led directly to the invention of the first accurate mechanical clocks. Galileo was also able to show that the period of oscillation of a simple pendulum is proportional to the square root of its length and does not depend on its mass.

A simple pendulum is sketched in figure. A bob of mass M is suspended by a massless cable or bar of length L from a point about which it pivots freely. The angle between the cable and the vertical is called θ. The force of gravity acting on the mass M, always equal to –Mg in the vertical direction, is a vector that may be resolved into two components, one that acts ineffectually along the cable and another, perpendicular to the cable that tends to restore the bob to its equilibrium position directly below the point of suspension. This latter component is given by,

$$F = M\,g \sin\theta$$

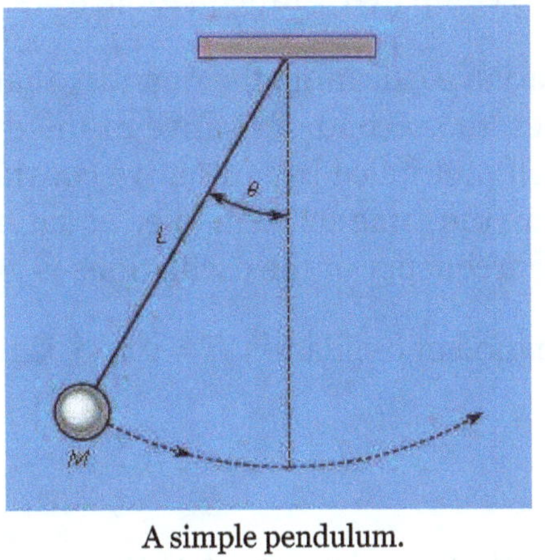

A simple pendulum.

The bob is constrained by the cable to swing through an arc that is actually a segment of a circle of radius L. If the cable is displaced through an angle θ, the bob moves a distance Lθ along its arc (θ must be expressed in radians for this form to be correct). Thus, Newton's second law may be written:

$$F = Mg = M\frac{d^2(L\theta)}{dt^2}.$$

Equating equation $F = Mg\sin\theta$ to equation $F = Mg = M\frac{d^2(L\theta)}{dt^2}$, one sees immediately that the mass M will drop out of the resulting equation. The simple pendulum is an example of a falling body, and its dynamics do not depend on its mass for exactly the same reason that the acceleration of a falling body does not depend on its mass: both the force of gravity and the inertia of the body are proportional to the same mass, and the effects cancel one another.

If the angle θ is sufficiently small, equation $\frac{d^2\theta}{dt^2} = -\frac{g}{L}\sin\theta$ may be rewritten in a form that is both more familiar and more amenable to solution. Figure shows a segment of a circle of radius L.

A radius vector at angle θ, as shown, locates a point on the circle displaced a distance Lθ along the arc. It is clear from the geometry that L sin θ and Lθ are very nearly equal for small θ. It follows then that sin θ and θ are also very nearly equal for small θ. Thus, if the analysis is restricted to small angles, then sin θ may be replaced by θ in equation $\frac{d^2\theta}{dt^2} = -\frac{g}{L}\sin\theta$ to obtain $\frac{d^2\theta}{dt^2} = -\frac{g}{L}\theta$.

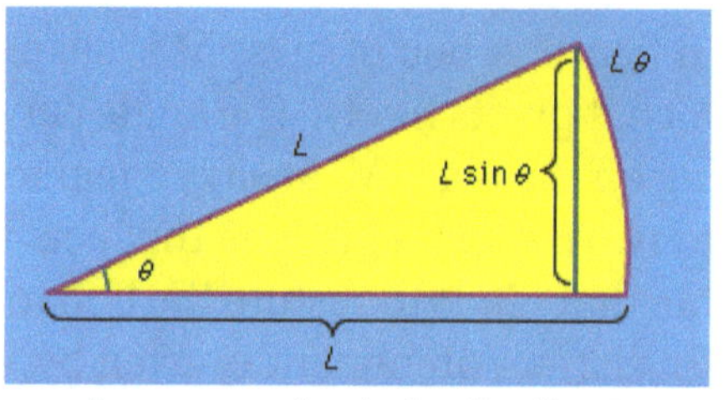

A segment of a circle of radius L.

Equation $\frac{d^2\theta}{dt^2} = -\frac{g}{L}\theta$ should be compared with equation $a = \frac{d^2x}{dt^2} = -\frac{k}{m}x$.: $d^2x/dt^2 = -(k/m)x$. In the first case, the dynamic variable (meaning the quantity that changes with time) is θ, in the second case it is x. In both cases, the second derivative of the dynamic variable with respect to time is equal to the variable itself multiplied by a negative constant. The equations are therefore mathematically identical and have the same solution—i.e., equation $x = A\cos\omega t.$, or θ = A cos ωt. In the case of the pendulum, the frequency of the oscillations is given by the constant in equation $\frac{d^2\theta}{dt^2} = -\frac{g}{L}\theta$, or $\omega^2 = g/L$. The period of oscillation, T = 2π/ω, is therefore,

$$T = 2\pi\sqrt{\frac{L}{g}}.$$

Just as Galileo concluded, the period is independent of the mass and proportional to the square root of the length.

As with most problems in physics, this discussion of the pendulum has involved a number of simplifications and approximations. Most obviously, sin θ was replaced by θ to obtain equation $\frac{d^2\theta}{dt^2} = -\frac{g}{L}\theta$. This approximation is surprisingly accurate. For example, at a not-very-small angle of 17.2°, corresponding to 0.300 radian, sin θ is equal to 0.296, an error of less than 2 percent. For smaller angles, of course, the error is appreciably smaller.

The problem was also treated as if all the mass of the pendulum were concentrated at a point at the end of the cable. This approximation assumes that the mass of the bob at the end of the cable is much larger than that of the cable and that the physical size of the bob is small compared with the length of the cable. When these approximations are not sufficient, one must take into account the way in which mass is distributed in the cable and bob. This is called the physical pendulum, as opposed to the idealized model of the simple pendulum. Significantly, the period of a physical pendulum does not depend on its total mass either.

The effects of friction, air resistance, and the like have also been ignored. These dissipative forces have the same effects on the pendulum as they do on any other kind of harmonic oscillator. They cause the amplitude of a freely swinging pendulum to grow smaller on successive swings. Conversely, in order to keep a pendulum clock going, a mechanism is needed to restore the energy lost to dissipative forces.

Circular Motion

Consider a particle moving along the perimeter of a circle at a uniform rate, such that it makes one complete revolution every hour. To describe the motion mathematically, a vector is constructed from the centre of the circle to the particle. The vector then makes one complete revolution every hour. In other words, the vector behaves exactly like the large hand on a wristwatch, an arrow of fixed length that makes one complete revolution every hour. The motion of the point of the vector is an example of uniform circular motion, and the period T of the motion is equal to one hour (T = 1 h). The arrow sweeps out an angle of 2π radians (one complete circle) per hour. This rate is called the angular frequency and is written $\omega = 2\pi\ h^{-1}$.

$$T = \frac{2\pi}{\omega}.$$

These definitions and relations are the same as they are for harmonic motion.

Consider a coordinate system, as shown in Figure, with the circle centred at the origin. At any instant of time, the position of the particle may be specified by giving the radius r of the circle and the angle θ between the position vector and the x-axis. Although r is constant, θ increases uniformly with time t, such that θ = ωt, or dθ/dt = ω, where ω is the angular frequency in equation $T = \frac{2\pi}{\omega}$. Contrary to the case of the wristwatch, however, ω is positive by convention when the rotation is in the counterclockwise sense. The vector r has x and y components in figure.

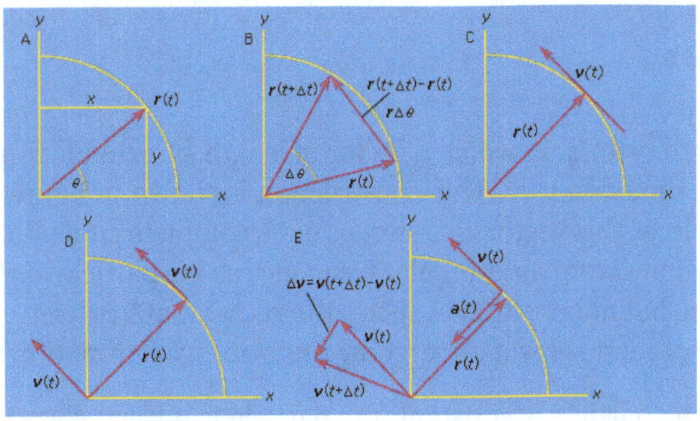

Figure: (A) A coordinate system to describe uniform circular motion. (B) The distance traveled in time Δt by a particle undergoing uniform circular motion. (C) The instantaneous velocity of the particle. (D) The velocity vector v undergoes uniform circular motion at the same angular frequency as the particle. (E) The acceleration vector of the particle.

One meaning of equations $x = r\cos\theta = r\cos\omega t$ and $y = r\sin\theta = r\sin\omega t$. is that, when a particle undergoes uniform circular motion, its x and y components each undergo simple harmonic motion. They are, however, not in phase with one another: at the instant when x has its maximum amplitude (say, at θ = 0), y has zero amplitude, and vice versa.

In a short time, Δt, the particle moves rΔθ along the circumference of the circle, as shown in figure . The average speed of the particle is thus given by,

$$\bar{v} = r\frac{\Delta\theta}{\Delta t}.$$

The average velocity of the particle is a vector given by,

$$\bar{v} = \frac{r(t+\Delta t) - r(t)}{\Delta t}.$$

This operation of vector subtraction is indicated in figure. It yields a vector that is nearly perpendicular to r(t) and r(t + Δt). Indeed, the instantaneous velocity, found by allowing Δt to shrink to zero, is a vector v that is perpendicular to r at every instant and whose magnitude is,

$$|v| = r\frac{d\theta}{dt} = r\omega.$$

The relationship between r and v is shown in figure. It means that the particle's instantaneous velocity is always tangent to the circle.

Notice that, just as the position vector r may be described in terms of the components x and y given by equations $x = r\cos\theta = r\cos\omega t$ and $y = r\sin\theta = r\sin\omega t$., the velocity vector v may be described in terms of its projections on the x and y axes, given by:

$$v_x = \frac{dx}{dt} = -r\omega\sin\omega t,$$

$$v_y = \frac{dy}{dt} = r\omega\cos\omega t.$$

Imagine a new coordinate system, in which a vector of length ωr extends from the origin and points at all times in the same direction as v. This construction is shown in figure. Each time the particle sweeps out a complete circle, this vector also sweeps out a complete circle. In fact, its point is executing uniform circular motion at the same angular frequency as the particle itself. Because vectors have magnitude and direction, but not position in space, the vector that has been constructed is the velocity v. The velocity of the particle is itself undergoing uniform circular motion at angular frequency ω.

Although the speed of the particle is constant, the particle is nevertheless accelerated, because its velocity is constantly changing direction. The acceleration a is given by,

$$a = \frac{dv}{dt}.$$

Since v is a vector of length rω undergoing uniform circular motion, equations $\bar{v} = r\frac{\Delta\theta}{\Delta t}$. and $\bar{v} = \frac{r(t+\Delta t)-r(t)}{\Delta t}$ may be repeated, as illustrated in figure, giving:

$$\bar{v} = r\omega\frac{\Delta\theta}{\Delta t}.$$

Thus, one may conclude that the instantaneous acceleration is always perpendicular to v and its magnitude is,

$$|a| = r\omega\frac{d\theta}{dt} = r\omega^2.$$

Since v is perpendicular to r, and a is perpendicular to v, the vector a is rotated 180° with respect to r. In other words, the acceleration is parallel to r but in the opposite direction. The same conclusion may be reached by realizing that a has x and y components given by similar to equations,

$$v_x = \frac{dx}{dt} = -r\omega\sin\omega t, \text{ and } v_y = \frac{dy}{dt} = r\omega\cos\omega t.$$

When equations $a_x = \frac{dv_x}{dt} = -r\omega^2\cos\omega t$, and $a_y = \frac{dv_y}{dt} = -r\omega^2\sin\omega t$, are compared with equations $x = r\cos\theta = r\cos\omega t$ and $y = r\sin\theta = r\sin\omega t$. for x and y, it is clear that the components of a are just those of r multiplied by $-\omega^2$, so that $a = -\omega^2 r$. This acceleration is called the centripetal acceleration, meaning that it is inward, pointing along the radius vector toward the centre of the circle. It is sometimes useful to express the centripetal acceleration in terms of the speed v. Using $v = \omega r$, one can write,

$$a = -\frac{v^2}{r}.$$

Circular Orbits

The detailed behaviour of real orbits is the concern of celestial mechanics.

A body in uniform circular motion undergoes at all times a centripetal acceleration given by equation $a = -\frac{v^2}{r}$. According to Newton's second law, a force is required to produce this acceleration. In the case of an orbiting planet, the force is gravity. The situation is illustrated in figure. The gravitational attraction of the Sun is an inward (centripetal) force acting on Earth. This force produces the centripetal acceleration of the orbital motion.

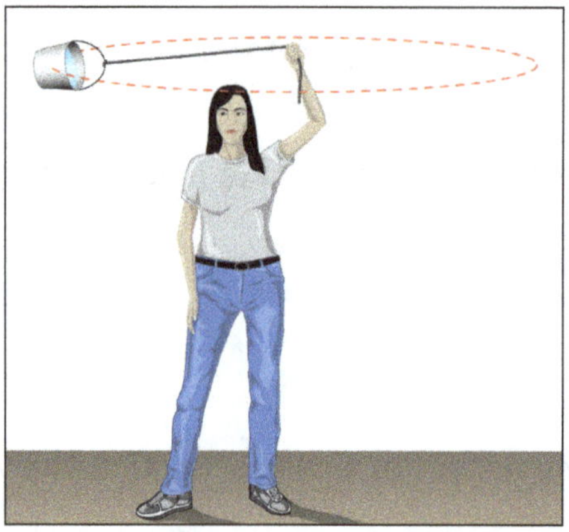

Centripetal force: The bucket experiences a centripetal force directed along the string toward the centre of the circle.

Before these ideas are expressed quantitatively, an understanding of why a force is needed to maintain a body in an orbit of constant speed is useful. The reason is that, at each instant, the velocity of the planet is tangent to the orbit. In the absence of gravity, the planet would obey the law of inertia (Newton's first law) and fly off in a straight line in the direction of the velocity at constant speed. The force of gravity serves to overcome the inertial tendency of the planet, thereby keeping it in orbit.

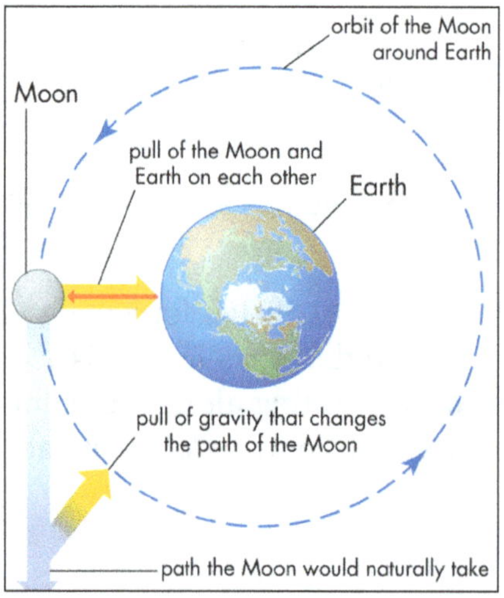

Effects of gravity on Earth and the Moon.

The gravitational force between two bodies such as the Sun and Earth is given by,

$$F = -G\frac{M_s M_E}{r^2},$$

where M_S and M_E are the masses of the Sun and Earth, respectively, r is the distance between their centres, and G is a universal constant equal to 6.674×10^{-11} Nm^2/kg^2 (Newton metres squared per kilogram squared). The force acts along the direction connecting the two bodies (i.e., along the radius vector of the uniform circular motion), and the minus sign signifies that the force is attractive, acting to pull Earth toward the Sun.

To an observer on the surface of Earth, the planet appears to be at rest at (approximately) a constant distance from the Sun. It would appear to the observer, therefore, that any force (such as the Sun's gravity) acting on Earth must be balanced by an equal and opposite force that keeps Earth in equilibrium. In other words, if gravity is trying to pull Earth into the Sun, some opposing force must be present to prevent that from happening. In reality, no such force exists. Earth is in freely accelerated motion caused by an unbalanced force. The apparent force, known in mechanics as a pseudoforce, is due to the fact that the observer is actually in accelerated motion. In the case of orbital motion, the outward pseudoforce that balances gravity is called the centrifugal force.

For a uniform circular orbit, gravity produces an inward acceleration given by equation $a = -\frac{v^2}{r}$. , $a = -v^2/r$. The pseudoforce f needed to balance this acceleration is just equal to the mass of Earth times an equal and opposite acceleration, or $f = M_E v^2/r$. The earthbound observer then believes that there is no net force acting on the planet—i.e., that F + f = 0, where F is the force of gravity given by equation $F = -G\frac{M_s M_E}{r^2}$,. Combining these equations yields a relation between the speed v of a planet and its distance r from the Sun:

$$v^2 = G\frac{M_s}{r},$$

It should be noted that the speed does not depend on the mass of the planet. This occurs for exactly the same reason that all bodies fall toward Earth with the same acceleration and that the period of a pendulum is independent of its mass. An orbiting planet is in fact a freely falling body.

Equation $v^2 = G\frac{M_s}{r}$, is a special case (for circular orbits) of Kepler's third law. Using the fact that $v = 2\pi r/T$, where $2\pi r$ is the circumference of the orbit and T is the time to make a complete orbit (i.e., T is one year in the life of the planet), it is easy to show that $T^2 = (4\pi^2/GMS)r^3$. This relation also may be applied to satellites in circular orbit around Earth (in which case, ME must be substituted for MS) or in orbit around any other central body.

Angular Momentum and Torque

A particle of mass m and velocity v has linear momentum p = mv. The particle may also have angular momentum L with respect to a given point in space. If r is the vector from the point to the particle then,

$$L = r \times P.$$

Notice that angular momentum is always a vector perpendicular to the plane defined by the vectors r and p (or v). For example, if the particle (or a planet) is in a circular orbit, its angular momentum with respect to the centre of the circle is perpendicular to the plane of the orbit and in the direction given by the vector cross product right-hand rule, as shown in Figure. Moreover, since in the case of a circular orbit, r is perpendicular top (or v), the magnitude of L is simply,

$$L = rp = mvr.$$

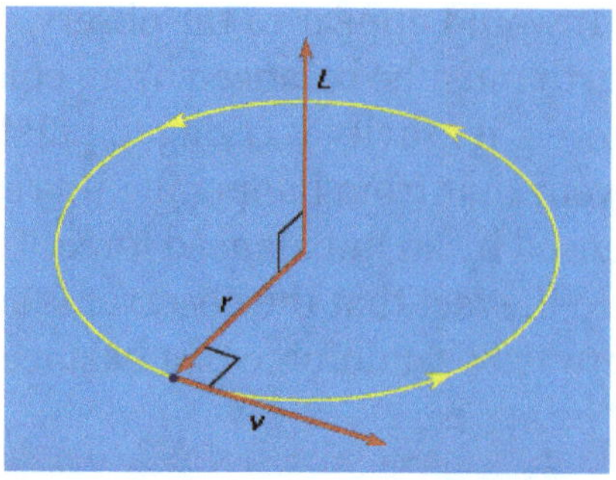

The angular momentum L of a particle traveling in a circular orbit.

The significance of angular momentum arises from its derivative with respect to time,

$$\frac{dL}{dt}=\frac{d}{dt}(r\times p)=m\frac{d}{dt}(r\times v).$$

where p has been replaced by mv and the constant m has been factored out. Using the product rule of differential calculus,

$$\frac{d}{dt}(r \times v)=\frac{dr}{dt}\times v + r \times \frac{dv}{dt}.$$

In the first term on the right-hand side of equation $\frac{d}{dt}(r \times v)=\frac{dr}{dt}\times v + r \times \frac{dv}{dt}$, dr/dt is simply the velocity v, leaving v × v. Since the cross product of any vector with itself is always zero, that term drops out leaving,

$$\frac{d}{dt}(r \times v)=r\times\frac{dv}{dt}.$$

Here, d**v**/dt is the acceleration a of the particle. Thus, if equation $\frac{d}{dt}(r \times v)=r\times\frac{dv}{dt}$ is multiplied by m, the left-hand side becomes d**L**/dt, as in equation $\frac{dL}{dt}=\frac{d}{dt}(r\times p)=m\frac{d}{dt}(r\times v)$, and the right-hand side may be written r × ma. Since, according to Newton's second law, ma is equal to F, the net force acting on the particle, the result is,

$$\frac{dL}{dt}=r \times F.$$

Equation $\frac{dL}{dt}=r \times F$ means that any change in the angular momentum of a particle must be produced by a force that is not acting along the same direction as r. One particularly important application is the solar system. Each planet is held in its orbit by its gravitational attraction to the Sun, a force that acts along the vector from the Sun to the planet. Thus, the force of gravity cannot change the angular momentum of any planet with respect to the Sun. Therefore, each planet has constant angular momentum with respect to the Sun. This conclusion is correct even though the real orbits of the planets are not circles but ellipses.

The quantity r × F is called the torque τ. Torque may be thought of as a kind of twisting force, the kind needed to tighten a bolt or to set a body into rotation. Using this definition, equation $\frac{dL}{dt} = r \times F$ may be rewritten as,

$$\tau = r \times F = \frac{dL}{dt}.$$

Equation $\tau = r \times F = \frac{dL}{dt}$ means that if there is no torque acting on a particle, its angular momentum is constant, or conserved. Suppose, however, that some agent applies a force F_a to the particle resulting in a torque equal to $r \times F_a$. According to Newton's third law, the particle must apply a force $-F_a$ to the agent. Thus, there is a torque equal to $-r \times F_a$ acting on the agent. The torque on the particle causes its angular momentum to change at a rate given by dL/dt = r × Fa. However, the angular momentum L_a of the agent is changing at the rate $dL_a/dt = -r \times F_a$. Therefore, $dL/dt + dL_a/dt = 0$, meaning that the total angular momentum of particle plus agent is constant, or conserved. This principle may be generalized to include all interactions between bodies of any kind, acting by way of forces of any kind. Total angular momentum is always conserved. The law of conservation of angular momentum is one of the most important principles in all of physics.

Motion of a Group of Particles

Centre of Mass

The word particle has been used to signify an object whose entire mass is concentrated at a point in space. In the real world, however, there are no particles of this kind. All real bodies have sizes and shapes. Furthermore, as Newton believed and is now known, all bodies are in fact compounded of smaller bodies called atoms. Therefore, the science of mechanics must deal not only with particles but also with more complex bodies that may be thought of as collections of particles.

To take a specific example, the orbit of a planet around the Sun was discussed earlier as if the planet and the Sun were each concentrated at a point in space. In reality, of course, each is a substantial body. However, because each is nearly spherical in shape, it turns out to be permissible, for the purposes of this problem, to treat each body as if its mass were concentrated at its centre. This is an example of an idea that is often useful in discussing bodies of all kinds: the centre of mass. The centre of mass of a uniform sphere is located at the centre of the sphere. For many purposes (such as the one cited above) the sphere may be treated as if all its mass were concentrated at its centre of mass.

To extend the idea farther, consider Earth and the Sun not as two separate bodies but as a single system of two bodies interacting with one another by means of the force of gravity. In the previous discussion of circular orbits, the Sun was assumed to be at rest at the centre of the orbit, but, according to Newton's third law, it must actually be accelerated by a force due to Earth that is equal and opposite to the force that the Sun exerts on Earth. In other words, considering only the Sun and Earth (ignoring, for example, all the other planets), if MS and ME are, respectively, the masses of the Sun and Earth, and if a_S and a_E are their respective accelerations, then combining Newton's second and third laws results in the equation $M_S a_S = -M_E a_E$. Writing each a as dv/dt, this equation is easily manipulated to give,

$$\frac{d}{dt}(M_S v_S + M_E v_E) = 0,$$

$$M_S v_S + M_E v_E = \text{constant}.$$

This remarkable result means that, as Earth orbits the Sun and the Sun moves in response to Earth's gravitational attraction, the entire two-body system has constant linear momentum, moving in a straight line at constant speed. Without any loss of generality, one can imagine observing the system from a frame of reference moving along with that same speed and direction. This is sometimes called the centre-of-mass frame. In this frame, the momentum of the two-body system—i.e., the constant in equation $M_S v_S + M_E v_E = \text{constant}$—is equal to zero.

Writing each of the v's as the corresponding d**r**/dt, equation $M_S v_S + M_E v_E = \text{constant}$ may be expressed in the form,

$$\frac{d}{dt}(M_S r_S + M_E r_E) = 0,$$

Thus, $M_S r_S$ and $M_E r_E$ are two vectors whose vector sum does not change with time. The sum is defined to be the constant vector MR, where M is the total mass of the system and equals $M_S + M_E$. Thus,

$$MR = M_S r_S + M_E r_E.$$

This procedure defines a constant vector R, from any arbitrarily chosen point in space. The relation between vectors R, r_S, and rE is shown in Figure. The fact that R is constant (although r_S and r_E are not constant) means that, rather than Earth orbiting the Sun, Earth and the Sun are both orbiting an imaginary point fixed in space. This point is known as the centre of mass of the two-body system.

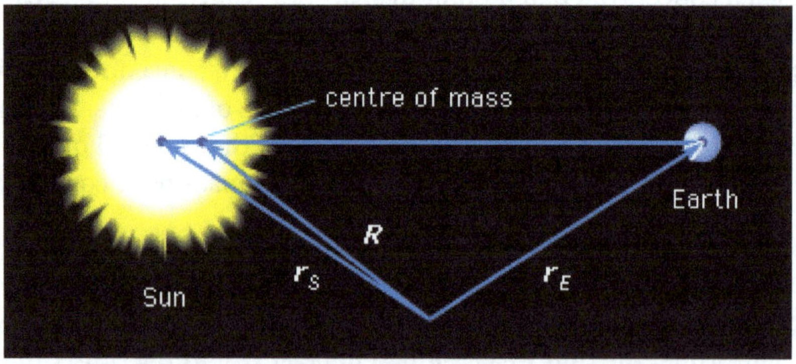

The centre of mass of the two-body Earth-Sun system.

Knowing the masses of the two bodies (MS = 1.99 × 10^{30} kilograms, ME = 5.98 × 10^{24} kilograms), it is easy to find the position of the centre of mass. The origin of the coordinate system may be chosen to be located at the centre of mass merely by defining R = 0. Then $r_S = (M_E/M_S)\, r_E \approx 450$ kilometres, when r_E is rounded to 1.5×10^8 km. A few hundred kilometres is so small compared to r_E that, for all practical purposes, no appreciable error occurs when r_S is ignored and the Sun is assumed to be stationary at the centre of the orbit.

With this example as a guide, it is now possible to define the centre of mass of any collection of bodies. Assume that there are N bodies altogether, each labeled with numbers ranging from 1 to N, and that the vector from an arbitrary origin to the ith body—where i is some number between 1 and N—is $\mathbf{r}_i$, as shown in Figure. Let the mass of the ith body be mi. Then the total mass of the N-body system is:

$$m = \sum_{i=1}^{N} m_i,$$

and the centre of mass of the system is found at the end of a vector R given by,

$$mR = \sum_{i=1}^{N} m_i r_i,$$

as illustrated in figure below. This definition applies regardless of whether the N bodies making up the system are the stars in a galaxy, the atoms in a rigid body, larger and arbitrarily chosen segments of a rigid body, or any other system of masses.

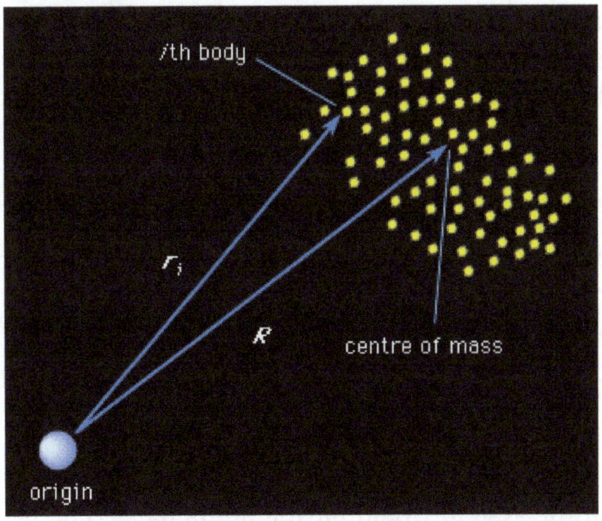

The centre of mass of an N-body system.

According to equation $mR = \sum_{i=1}^{N} m_i r_i$, the vector to the centre of mass of any system is a kind of weighted average of the vectors to all the components of the system.

The statics and dynamics of many complicated bodies or systems may often be understood by simply applying Newton's laws as if the system's mass were concentrated at the centre of mass.

Conservation of Momentum

Newton's second law, in its most general form, says that the rate of a change of a particle's momentum p is given by the force acting on the particle; i.e., F = dp/dt. If there is no force acting on the particle, then, since dp/dt = 0, p must be constant, or conserved. This observation is merely a restatement of Newton's first law, the principle of inertia: if there is no force acting on a body, it moves at constant speed in a straight line.

Now suppose that an external agent applies a force F_a to the particle so that p changes according to,

$$\frac{dp}{dt} = F_a.$$

According to Newton's third law, the particle must apply an equal and opposite force $-F_a$ to the external agent. The momentum p_a of the external agent therefore changes according to

$$\frac{dpa}{dt} = -F_a.$$

Adding together equations $\frac{dp}{dt} = F_a$ and $\frac{dpa}{dt} = -F_a$ results in the equation,

$$\frac{d}{dt}(p + pa) = 0.$$

The force applied by the external agent changes the momentum of the particle, but at the same time the momentum of the external agent must also change in such a way that the total momentum of both together is constant, or conserved. This idea may be generalized to give the law of conservation of momentum: in all the interactions between all the bodies in the universe, total momentum is always conserved.

It is useful in this light to examine the behaviour of a complicated system of many parts. The centre of mass of the system may be found using equation $mR = \sum_{i=1}^{N} m_i r_i$,. Differentiating with respect to time gives,

$$mv = \sum_{i=1}^{N} m_i v_i$$

where **v** = d*R*/dt and $v_i = dr_i/dt$. Note that $m_i v_i$ is the momentum of the ith part of the system, and mv is the momentum that the system would have if all its mass (i.e., m) were concentrated at its centre of mass, the point whose velocity is v. Thus, the momentum associated with the centre of mass is the sum of the momenta of the parts.

Suppose now that there is no external agent applying a force to the entire system. Then the only forces acting on the system are those exerted by the parts on one another. These forces may accelerate the individual parts. Differentiating equation $mv = \sum_{i=1}^{N} m_i v_i$ with respect to time gives,

$$m\frac{dv}{dt} = \sum_{i=1}^{N} m_i \frac{dv_i}{dt} = \sum_{i=1}^{N} F_i.$$

where F_i is the net force, or the sum of the forces, exerted by all the other parts of the body on the ith part. F_i is defined mathematically by the equation,

$$F_i = \sum_{j=1}^{N} F_{ij},$$

where F_{ij} represents the force on body i due to body j (the force on body i due to itself, F_{ii}, is zero). The motion of the centre of mass is then given by the complicated-looking formula,

$$m\frac{dv}{dt} = \sum_{i=1}^{N}\left(\sum_{j=1}^{N} F_{ij}\right).$$

This complicated formula may be greatly simplified, however, by noting that Newton's third law requires that for every force $\mathbf{F}_{ij}$ exerted by the jth body on the ith body, there is an equal and opposite force $-F_{ij}$ exerted by the ith body on the jth body. In other words, every term in the double sum has an equal and opposite term. The double summation on the right-hand side of equation $F_i = \sum_{j=1}^{N} F_{ij}$, always adds up to zero. This result is true regardless of the complexity of the system, the nature of the forces acting between the parts, or the motions of the parts. In short, in the absence of external forces acting on the system as a whole, $m d\mathbf{v}/dt = 0$, which means that the momentum of the centre of mass of the system is always conserved. Having determined that momentum is conserved whether or not there is an external force acting, one may conclude that the total momentum of the universe is always conserved.

Collisions

A collision is an encounter between two bodies that alters at least one of their courses. Altering the course of a body requires that a force be applied to it. Thus, each body exerts a force on the other. These forces of interaction may operate at some distance, as do the gravitational and electromagnetic forces, or the bodies may appear to make physical contact. However, even apparent contact between two bodies is only a macroscopic manifestation of microscopic forces that act between atoms some distance apart. There is no fundamental distinction between physical contact and interaction at a distance.

The importance of understanding the mechanics of collisions is obvious to anyone who has ever driven an automobile. In modern physics, however, collisions are important for a different reason. The current understanding of the subatomic particles of which atoms are composed is derived entirely from studying the results of collisions among them. Thus, in modern physics, the description of collisions is a significant part of the understanding of matter. These descriptions are quantum mechanical rather than classical, but they are nevertheless closely based on principles that arise out of classical mechanics.

It is possible in principle to predict the result of a collision using Newton's second law directly. Suppose that two bodies are going to collide and that F, the force of interaction between them, is known to be a function of r, the distance between them. Then, if it is known that, say, one particle has incident momentum p, the problem is solved if the final momentum p + Δp can be determined. Inverting Newton's second law, F = dp/dt, the change in momentum is given by,

$$\Delta p = \int_{-\infty}^{\infty} F dt.$$

This integral is known as the impulse imparted to the particle. In order to perform the integral, it is necessary to know r at all times so that F may be known at all times. More realistically, Δp is the sum of a series of small steps such that,

$$\delta p = F\delta t,$$

where F depends on the instantaneous distance between the particles. Because p = mv = mdr/dt, the change in r in this step is,

$$\delta r = \frac{p}{m}\delta t.$$

At the next step, there is a new distance, r + δr, giving a new value of the force in equation and a new momentum, p + δp, in equation $\delta r = \frac{p}{m}\delta t$. This method of analyzing collisions is used in numerical calculations on digital computers.

To predict the result of a collision analytically (rather than numerically) it is often most useful to apply conservation laws. In any collision (as in any other phenomenon), energy, momentum, and angular momentum are always conserved. Judicious application of these laws may be extremely useful because they do not depend in any way on the detailed nature of the interaction (i.e., the force as a function of distance).

This point can be illustrated by the following example. A collision is to take place between two bodies of the same mass m. One of the bodies is initially at rest (its momentum is zero). The other has initial momentum p_0. After the collision, the body previously at rest has momentum p_1, and the body initially in motion has momentum p_2. Since momentum is conserved, the total momentum after the collision, $p_1 + p_2$, must be equal to the total momentum before the collision, p_0; that is,

$$P_0 = P_1 + P_2.$$

Equation $P_0 = P_1 + P_2$ is the equation of a vector triangle, as shown in figure. However, p_1 and p_2 are not determined by this condition; they are only constrained by it.

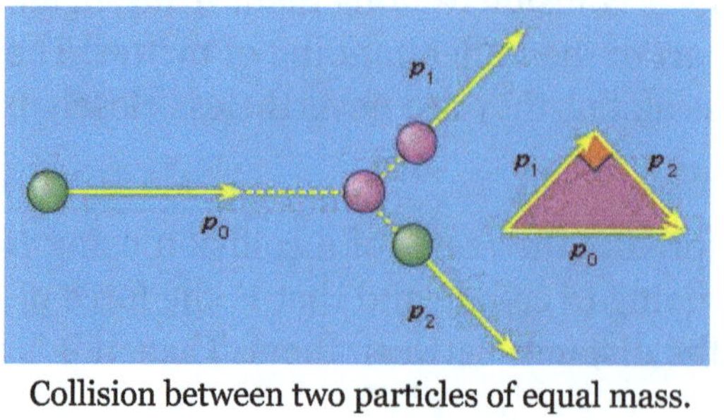

Collision between two particles of equal mass.

Although energy is always conserved, the kinetic energy of the incident body is not always converted entirely into the kinetic energy of the two bodies after the collision. For example, if the bodies are microscopic (say, two identical atoms), the collision may cause one or both to be excited into a state of higher internal energy than it started with. Such an event would leave correspondingly less kinetic energy for the outgoing atoms. In fact, it is precisely by studying the trajectories of outgoing projectiles in collisions like these that physicists are able to determine the possible excited states of microscopic particles.

In a collision between macroscopic objects, some of the kinetic energy is always converted to heat. Heat is the energy of random vibrations of the atoms and molecules that constitute the bodies. However, if the amount of heat is negligible compared to the initial kinetic energy, it may be ignored. Such a collision is said to be elastic.

Suppose the collision described above between two bodies, each of mass m, is between billiard balls, and suppose it is elastic (a reasonably good approximation of real billiard balls). The kinetic energy of the incident ball is then equal to the sum of the kinetic energies of the outgoing balls. According to equation $K = \frac{1}{2}mv^2$, the kinetic energy of a moving object is given by K = 1/2mv2, where v is the speed of the ball (technically, the energy associated with the fact that the ball is rolling as well as translating is ignored here; see below Rotation about a moving axis). Equation $K = \frac{1}{2}mv^2$ may be written in a particularly useful form by recognizing that since p = mv,

$$K = \frac{1}{2}mv^2 = \frac{p^2}{2m}.$$

Then the conservation of kinetic energy may be written,

$$\frac{p_0^2}{2m} = \frac{p_1^2}{2m} + \frac{p_2^2}{2m},$$

or, canceling the factors 2m,

$$p_0^2 = p_1^2 + p_2^2$$

Comparing this result with equation $P_0 = P_1 + P_2$ shows that the vector triangle is pythagorean; p_1 and p_2 are perpendicular. This result is well known to all experienced pool players. Notice that it was possible to arrive at this result without any knowledge of the forces that act when billiard balls collide.

Relative Motion

A collision between two bodies can always be described in a frame of reference in which the total momentum is zero. This is the centre-of-mass (or centre-of-momentum) frame mentioned earlier. Then, for example, in the collision between two bodies of the same mass discussed above, the two bodies always have equal and opposite velocities, as shown in Figure. It should be noted that, in this frame of reference, the outgoing momenta are antiparallel and not perpendicular.

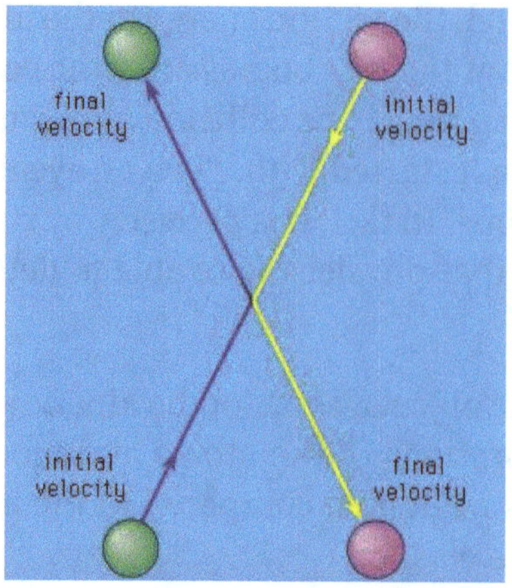

Collision between two particles of equal mass as seen from the centre-of-mass frame of reference.

Any collection of bodies may similarly be described in a frame of reference in which the total momentum is zero. This frame is simply the one in which the centre of mass is at rest. This fact is easily seen by differentiating equation $mR = \sum_{i=1}^{N} m_i r_i$, with respect to time giving,

$$m\frac{dR}{dt} = \sum_{i=1}^{N} m_i \frac{dr_i}{dt}.$$

The right-hand side is the sum of the momenta of all the bodies. It is equal to zero if the velocity of the centre of mass, dR/dt, is equal to zero.

If Newton's second law is correct in any frame of reference, it will also appear to be correct to an observer moving with any constant velocity with respect to that frame. This principle, called the principle of Galilean relativity, is true because, to the moving observer, the same constant velocity seems to have been added to the velocity of every particle in the system. This change does not affect the accelerations of the particles (since the added velocity is constant, not accelerated) and therefore does not change the apparent force (mass times acceleration) acting on each particle. That is why it is permissible to describe a problem from the centre-of-momentum frame (provided that the centre of mass is not accelerated) or from any other frame moving at constant velocity with respect to it.

If this principle is strictly correct, the fundamental forces of physics should not contain any particular speed. This must be true because the speed of any object will be different to observers in different but equally good frames of reference, but the force should always be the same. It turns out, according to the theory of James Clerk Maxwell, that there is an intrinsic speed in the force laws of electricity and magnetism: the speed of light appears in the forces between electric charges and between magnetic poles. This discrepancy was ultimately resolved by Albert Einstein's special theory of relativity. According to the special theory of relativity, Newtonian mechanics breaks down when the relative speed between particles approaches the speed of light.

Coupled Oscillators

In the section on simple harmonic oscillators, the motion of a single particle held in place by springs was considered. In this section, the motion of a group of particles bound by springs to one another is discussed. The solutions of this seemingly academic problem have far-reaching implications in many fields of physics. For example, a system of particles held together by springs turns out to be a useful model of the behaviour of atoms mutually bound in a crystalline solid.

To begin with a simple case, consider two particles in a line, as shown in Figure. Each particle has mass m, each spring has spring constant k, and motion is restricted to the horizontal, or x, direction. Even this elementary system is capable of surprising behaviour, however. For instance, if one particle is held in place while the other is displaced, and then both are released, the displaced particle immediately begins to execute simple harmonic motion. This motion, by stretching the spring between the particles, starts to excite the second particle into motion. Gradually the energy of motion passes from the first particle to the second until a point is reached at which the first particle is at rest and only the second is oscillating. Then the process starts all over again, the energy passing in the opposite direction.

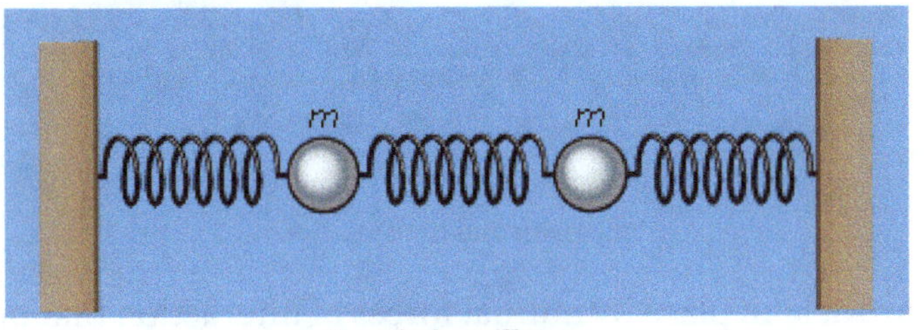

Coupled oscillators.

To analyze the possible motions of the system, one writes equations similar to equation $a = \frac{d^2x}{dt^2} = -\frac{k}{m}x$, giving the acceleration of each particle owing to the forces acting on it. There is one equation for each particle (two equations in this case). The force on each particle depends not only on its displacement from its equilibrium position but also on its distance from the other particle, since the spring between them stretches or compresses according to that distance. For this reason the motions are coupled, the solution of each equation (the motion of each particle) depending on the solution of the other (the motion of the other).

Analyzing the system yields the fact that there are two special states of motion in which both particles are always in oscillation with the same frequency. In one state, the two particles oscillate in opposite directions with equal and opposite displacements from equilibrium at all times. In the other state, both particles move together, so that the spring between them is never stretched or compressed. The first of these motions has higher frequency than the second because the centre spring contributes an increase in the restoring force.

These two collective motions, at different, definite frequencies, are known as the normal modes of the system.

If a third particle is inserted into the system together with another spring, there will be three equations to solve, and the result will be three normal modes. A large number N of particles in a line will have N normal modes. Each normal mode has a definite frequency at which all the particles oscillate. In the highest frequency mode each particle moves in the direction opposite to both of its neighbours. In the lowest frequency mode, neighbours move almost together, barely disturbing the springs between them. Starting from one end, the amplitude of the motion gradually builds up, each particle moving a bit more than the one before, reaching a maximum at the centre, and then decreasing again. A plot of the amplitudes, shown in Figure, basically describes one-half of a sine wave from one end of the system to the other. The next mode is a full sine wave, then 3/2 of a sine wave, and so on to the highest frequency mode, which may be visualized as N/2 sine waves. If the vibrations were up and down rather than side to side, these modes would be identical to the fundamental and harmonic vibrations excited by plucking a guitar string.

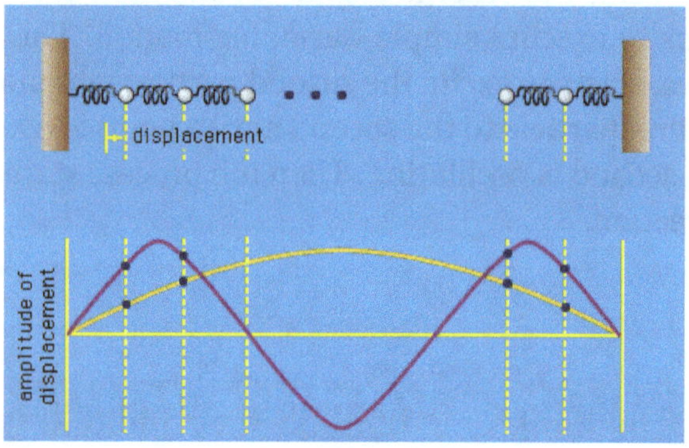

Normal modes.

The atoms of a crystal are held in place by mutual forces of interaction that oppose any disturbance from equilibrium positions, just as the spring forces in the example above. For small displacements of the atoms, they behave mathematically just like spring forces i.e., they obey Hooke's law, equation $F = -kx$. Each atom is free to move in three dimensions rather than one, however; therefore each atom added to a crystal adds three normal modes. In a typical crystal at ordinary temperature, all these modes are always excited by random thermal energy. The lower-frequency, longer-wavelength modes may also be excited mechanically. These are called sound waves.

Rigid Body

A rigid body is an extended area of material that includes all the points inside it, and which moves so that the distances and angles between all its points remain constant. The location of a rigid body can be described by the position of one point P inside it, together with the rotation angle of the body (one angle in 2D, three angles in 3D).

	Location Description	Velocity Description
Point mass	Position vector $\vec{r}$ p	Velocity vector $\vec{v}$ p
Rigid body in 2D	Position vector $\vec{r}$ P angle θ	Velocity vector $\vec{v}$ P angular velocity ω
Rigid body in 3D	Position vector $\vec{r}$ P angles θ, ϕ, ψ	Velocity vector $\vec{v}$ P angular velocity vector $\vec{\omega}$

Neither point masses nor rigid bodies can physically exist, as no body can really be a single point with no extent, and no extended body can be exactly rigid. Despite this, these are very useful models for mechanics and dynamics.

Statics

Statics is the study of bodies and structures that are in equilibrium. For a body to be in equilibrium, there must be no net force acting on it. In addition, there must be no net torque acting on it. Figure below shows a body in equilibrium under the action of equal and opposite forces. Figure below shows a body acted on by equal and opposite forces that produce a net torque, tending to start it rotating. It is therefore not in equilibrium.

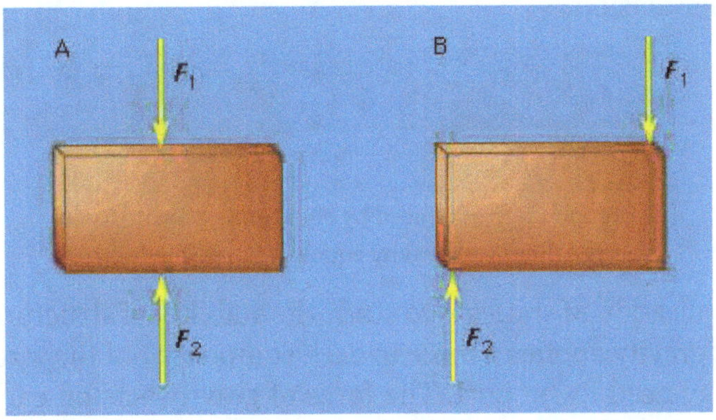

(A) A body in equilibrium under equal and opposite forces.
(B) A body not in equilibrium under equal and opposite forces.

When a body has a net force and a net torque acting on it owing to a combination of forces, all the forces acting on the body may be replaced by a single (imaginary) force called the resultant, which acts at a single point on the body, producing the same net force and the same net torque. The body can be brought into equilibrium by applying to it a real force at the same point, equal and opposite to the resultant. This force is called the equilibrant. An example is shown in the figure below.

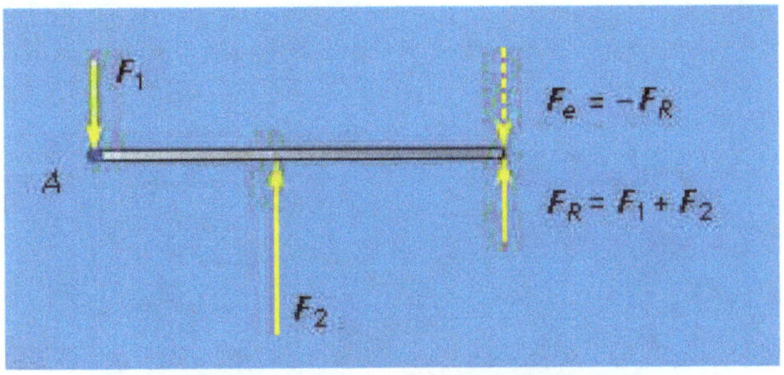

The resultant force (F_R) produces the same net force and the same net torque about point A as $F_1 + F_2$; the body can be brought into equilibrium by applying the equilibrant force F_e.

The torque on a body due to a given force depends on the reference point chosen, since the torque τ by definition equals r × F, where r is a vector from some chosen reference point to the point of application of the force. Thus, for a body to be at equilibrium, not only must the net force on it be equal to zero but the net torque with respect to any point must also be zero. Fortunately, it is easily shown for a rigid body that, if the net force is zero and the net torque is zero with respect to any one point, then the net torque is also zero with respect to any other point in the frame of reference.

A body is formally regarded as rigid if the distance between any set of two points in it is always constant. In reality no body is perfectly rigid. When equal and opposite forces are applied to a body, it is always deformed slightly. The body's own tendency to restore the deformation has the effect of applying counterforces to whatever is applying the forces, thus obeying Newton's third law. Calling a body rigid means that the changes in the dimensions of the body are small enough to be neglected, even though the force produced by the deformation may not be neglected.

Equal and opposite forces acting on a rigid body may act so as to compress the body or to stretch it. The bodies are then said to be under compression or under tension, respectively. Strings, chains, and cables are rigid under tension but may collapse under compression. On the other hand, certain building materials, such as brick and mortar, stone, or concrete, tend to be strong under compression but very weak under tension.

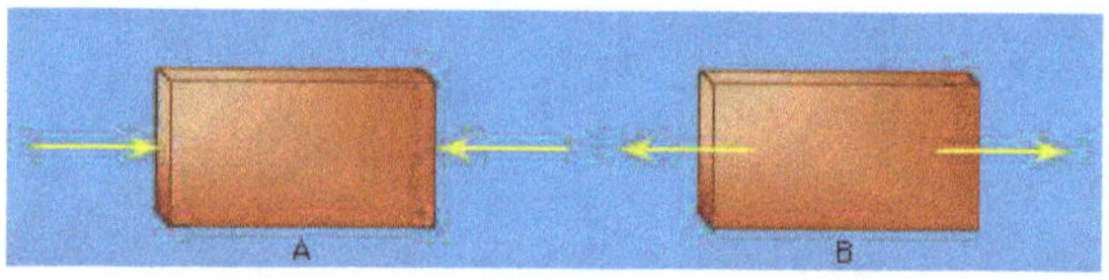

(A) Compression produced by equal and opposite forces.
(B) Tension produced by equal and opposite forces.

The most important application of statics is to study the stability of structures, such as edifices and bridges. In these cases, gravity applies a force to each component of the structure as well as to any bodies the structure may need to support. The force of gravity acts on each bit of mass of which each component is made, but for each rigid component it may be thought of as acting at a single point, the centre of gravity, which is in these cases the same as the centre of mass.

To give a simple but important example of the application of statics, consider the two situations shown in Figure. In each case, a mass m is supported by two symmetric members, each making an angle θ with respect to the horizontal. In Figure below the members are under tension; in Figure they are under compression. In either case, the force acting along each of the members is shown to be,

$$F = \frac{mg}{2\sin\theta}.$$

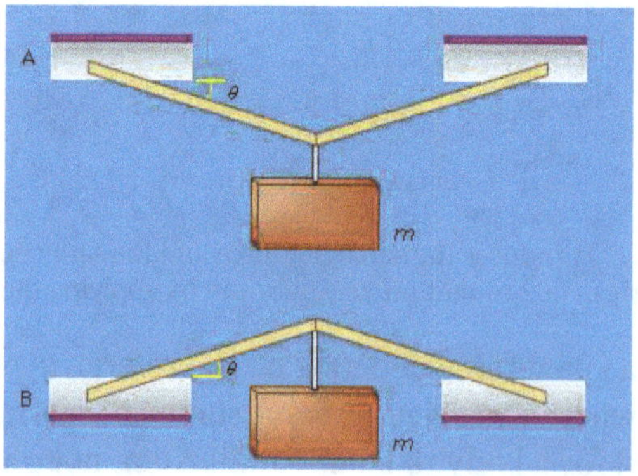

(A) A body supported by two rigid members under tension.
(B) A body supported by two rigid members under compression.

The force in either case thus becomes intolerably large if the angle θ is allowed to be very small.

In other words, the mass cannot be hung from thin horizontal members only capable of carrying either the compression or the tension forces of the mass.

The ancient Greeks built magnificent stone temples; however, the horizontal stone slabs that constituted the roofs of the temples could not support even their own weight over more than a very small span. For this reason, one characteristic that identifies a Greek temple is the many closely spaced pillars needed to hold up the flat roof. The problem posed by equation $F = \frac{mg}{2\sin\theta}$ was solved by the ancient Romans, who incorporated into their architecture the arch, a structure that supports its weight by compression, corresponding to the figure.

A suspension bridge illustrates the use of tension. The weight of the span and any traffic on it is supported by cables, which are placed under tension by the weight. Corresponding to Figure, the cables are not stretched to be horizontal, but rather they are always hung so as to have substantial curvature.

It should be mentioned in passing that equilibrium under static forces is not sufficient to guarantee the stability of a structure. It must also be stable against perturbations such as the additional forces that might be imposed, for example, by winds or by earthquakes. Analysis of the stability of structures under such perturbations is an important part of the job of an engineer or architect.

Rotation about a Fixed Axis

Consider a rigid body that is free to rotate about an axis fixed in space. Because of the body's inertia, it resists being set into rotational motion, and equally important, once rotating, it resists being brought to rest. Exactly how that inertial resistance depends on the mass and geometry of the body is discussed here.

Take the axis of rotation to be the z-axis. A vector in the x-y plane from the axis to a bit of mass fixed in the body makes an angle θ with respect to the x-axis. If the body is rotating, θ changes with time, and the body's angular frequency is,

$$\omega = \frac{d\theta}{dt},$$

ω is also known as the angular velocity. If ω is changing in time, there is also an angular acceleration α, such that:

$$\alpha = \frac{d\omega}{dt},$$

Because linear momentum p is related to linear speed v by p = mv, where m is the mass, and because force F is related to acceleration a by F = ma, it is reasonable to assume that there exists a quantity I that expresses the rotational inertia of the rigid body in analogy to the way m expresses the inertial resistance to changes in linear motion. One would expect to find that the angular momentum is given by:

$$L = I\omega$$

and that the torque (twisting force) is given by,

$$\tau = I\alpha.$$

One can imagine dividing the rigid body into bits of mass labeled m_1, m_2, m_3, and so on. Let the bit of mass at the tip of the vector be called m_i, as indicated in Figure. If the length of the vector from the axis to this bit of mass is R_i, then *mi*'s linear velocity *vi* equals ωR_i, and its angular momentum *Li* equals $m_i v i R_i$, or $m_i R_i^2 \omega$. The angular momentum of the rigid body is found by summing all the contributions from all the bits of mass labeled i = 1, 2, 3.

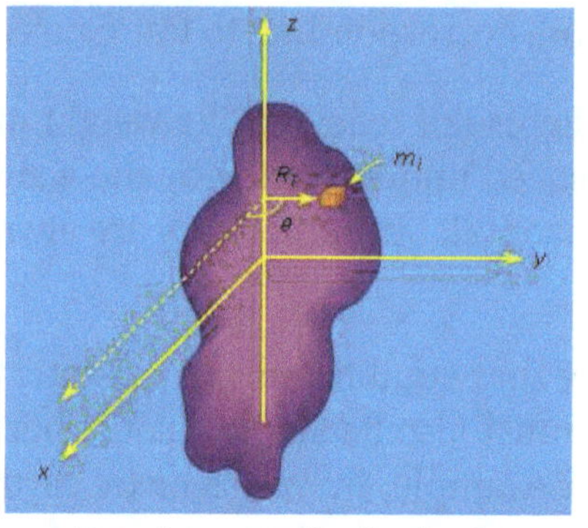

Rotation around a fixed axis.

In a rigid body, the quantity in parentheses in equation $L = \left(\sum_i m_i R_i^2\right)\omega.$ is always constant (each bit of mass *mi* always remains the same distance *Ri* from the axis). Thus if the motion is accelerated, then,

$$\frac{dL}{dt} = \left(\sum_i m_i R_i^2\right)\frac{d\omega}{dt}.$$

Recalling that τ = dL/dt, one may write,

$$\tau \quad \left(\sum m_i R_i\right)\alpha$$

(These equations may be written in scalar form, since L and τ are always directed along the axis of rotation in this topic.) Comparing equations $L = \left(\sum_i m_i R_i^2\right)\omega.$ and $\tau = \left(\sum_i m_i R_i^2\right)\alpha$ with $L = I\omega$ and $\tau = I\alpha$, one finds that,

$$I = \sum_i m_i R_i^2$$

The quantity *I* is called the moment of inertia.

According to equation $I = \sum_i m_i R_i^2$, the effect of a bit of mass on the moment of inertia depends on its distance from the axis. Because of the factor R_i^2, mass far from the axis makes a bigger contribution than mass close to the axis. It is important to note that R_i is the distance from the axis, not from a point. Thus, if x_i and y_i are the x and y coordinates of the mass mi, then $R_i^2 = x_i^2 + y_i^2$, regardless of the value of the z coordinate. The moments of inertia of some simple uniform bodies are given in the table.

Moments of Inertia for Uniform Bodies		
Body	axis	I
Thin rod (length L)	perpendicular axis through centre	$1/12\ ML^2$
Thin ring (radius R)	perpendicular axis through centre	MR^2
Solid circular cylinder	axis of cylinder	$1/12\ MR^2$
Thin disk	transverse axis through centre	$1/4\ MR^2$
Solid sphere	any axis through centre	$2/5\ MR^2$
Thin spherical shell	any axis through centre	$2/3\ MR^2$
Rectangular plate (length a, height b)	axis through centre perpendicular to the plate	$1/12\ M(a^2+b^2)$

The moment of inertia of any body depends on the axis of rotation. Depending on the symmetry of the body, there may be as many as three different moments of inertia about mutually perpendicular axes passing through the centre of mass. If the axis does not pass through the centre of mass, the moment of inertia may be related to that about a parallel axis that does so. Let I_c be the moment of inertia about the parallel axis through the centre of mass, r the distance between the two axes, and M the total mass of the body. Then,

$$I = I_c + Mr^2.$$

In other words, the moment of inertia about an axis that does not pass through the centre of mass is equal to the moment of inertia for rotation about an axis through the centre of mass (I_c) plus a contribution that acts as if the mass were concentrated at the centre of mass, which then rotates about the axis of rotation.

The dynamics of rigid bodies rotating about fixed axes may be summarized in three equations. The angular momentum is $L = I\omega$, the torque is $\tau = I\alpha$, and the kinetic energy is $K = {}^1/_2 I\omega^2$.

Rotation about a Moving Axis

The general motion of a rigid body tumbling through space may be described as a combination of translation of the body's centre of mass and rotation about an axis through the centre of mass. The linear momentum of the body of mass M is given by,

$$P = Mv_c,$$

where v_c is the velocity of the centre of mass. Any change in the momentum is governed by Newton's second law, which states that,

$$F = \frac{dp}{dt},$$

where F is the net force acting on the body. The angular momentum of the body with respect to any reference point may be written as,

$$L = L_c + r \times p,$$

where L_c is the angular momentum of rotation about an axis through the centre of mass, r is a vector from the reference point to the centre of mass, and $r \times p$ is therefore the angular momentum associated with motion of the centre of mass, acting as if all the body's mass were concentrated at that point. The quantity L_c in equation $L = L_c + r \times p$, is sometimes called the body's spin, and $r \times p$ is called the orbital angular momentum. Any change in the angular momentum of the body is given by the torque equation,

$$\tau = \frac{dL}{dt}.$$

An example of a body that undergoes both translational and rotational motion is the Earth, which rotates about an axis through its centre once per day while executing an orbit around the Sun once per year. Because the Sun exerts no torque on the Earth with respect to its own centre, the orbital angular momentum of the Earth is constant in time. However, the Sun does exert a small torque on the Earth with respect to the planet's centre, owing to the fact that the Earth is not perfectly spherical. The result is a slow shifting of the Earth's axis of rotation, known as the precession of the equinoxes.

The kinetic energy of a body that is both translating and rotating is given by,

$$K = \frac{1}{2} M v_c^2 + \frac{1}{2} I \omega^2,$$

where I is the moment of inertia and ω is the angular velocity of rotation about the axis through the centre of mass.

A common example of combined rotation and translation is rolling motion, as exhibited by a billiard ball rolling on a table, or a ball or cylinder rolling down an inclined plane. Consider the latter example, illustrated in Figure. Motion is impelled by the force of gravity, which may be resolved into two components, F_N, which is normal to the plane, and F_p, which is parallel to it. In addition to gravity, friction plays an essential role. The force of friction, written as f, acts parallel to the plane, in opposition to the direction of motion, at the point of contact between the plane and the rolling body. If f is very small, the body will slide without rolling. If f is very large, it will prevent motion from occurring. The magnitude of f depends on the smoothness and composition of the body and the plane, and it is proportional to F_N, the normal component of the force.

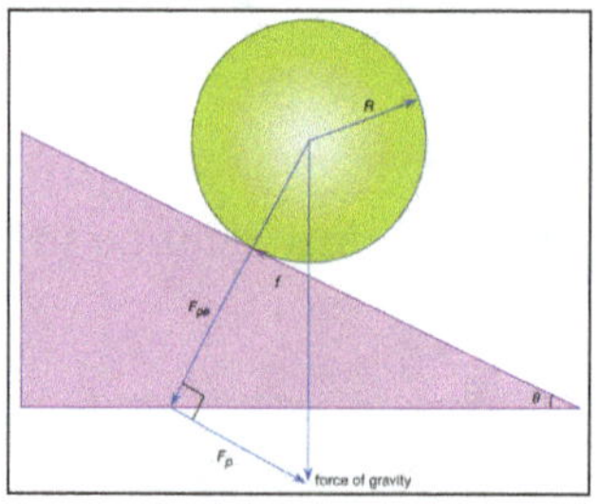

Rolling motion.

Consider a case in which *f* is just large enough to cause the body (sphere or cylinder) to roll without slipping. The motion may be analyzed from the point of view of an axis passing through the point of contact between the rolling body and the plane. Remarkably, the point of contact may always be regarded to be instantaneously at rest. To understand why, suppose that the rolling body has radius R and angular velocity ω about its centre-of-mass axis. Then, with respect to its own axis, each point on the circular crosssection in figure moves with instantaneous tangential linear speed $v_c = R\omega$. In particular, the point of contact is moving backward with this speed relative to the centre of mass. But with respect to the inclined plane, the centre of mass is moving forward with exactly this same speed. The net effect of the two equal and opposite speeds is that the point of contact is always instantaneously at rest. Therefore, although friction acts at that point, no work is done by friction, so mechanical energy (potential plus kinetic) may be regarded as conserved.

With respect to the axis through the point of contact, the torque is equal to RF_p, giving rise to an angular acceleration α given by $I_p\alpha = RF_p$, where I_p is the moment of inertia about the point-of-contact axis and can be determined by applying equation $I = I_c + Mr^2$. relating moments of inertia about parallel axes ($I_p = I + MR^2$). Thus,

$$\alpha = \frac{RF_p}{1 + MR^2}.$$

From this result, the motion of the body is easily obtained using the fact that the velocity of the centre of mass is $v_c = R\omega$ and hence the linear acceleration of the centre of mass is $a_c = R\alpha$.

Notice that, although without friction no angular acceleration would occur, the force of friction does not affect the magnitude of α. Because friction does no work, this same result may be obtained by applying energy conservation. The situation also may be analyzed entirely from the point of view of the centre of mass. In that case, the torque is $-fR$, but *f* also provides a linear force on the body. The *f* may then be eliminated by using Newton's second law and the fact that the torque equals the moment of inertia times the angular acceleration, once again leading to the same result.

One more interesting fact is hidden in the form of equation $\alpha = \frac{RF_p}{1 + MR^2}$. The parallel component of the force of gravity is given by,

$$F_p = Mg \sin\theta,$$

where θ is the angle of inclination of the plane. The moment of inertia about the centre of mass of any body of mass M may be written,

$$I = Mk^2,$$

where k is a distance called the radius of gyration. Comparison to equation $I = \sum_i m_i R_i^2$ shows that k is a measure of how far from the centre of mass the mass of the body is concentrated. Using equations $F_p = Mg \sin\theta$, and $I \quad Mk$ in equation $\alpha = \frac{RF_p}{1 + MR^2}$, one finds that:

$$\alpha = \frac{Rg \sin\theta}{k^2 + R^2}.$$

Thus, the angular acceleration of a body rolling down a plane does not depend on its total mass, although it does depend on its shape and distribution of mass. The same may be said of a_c, the linear acceleration of the centre of mass. The acceleration of a rolling ball, like the acceleration of a freely falling object, is independent of its mass. This observation helps to explain why Galileo was able to discover many of the basic laws of dynamics in gravity by studying the behaviour of balls rolling down inclined planes.

Motion in a Rotating Frame

Centrifugal Force

According to the principle of Galilean relativity, if Newton's laws are true in any reference frame, they are also true in any other frame moving at constant velocity with respect to the first one. Conversely, they do not appear to be true in any frame accelerated with respect to the first. Instead, in an accelerated frame, objects appear to have forces acting on them that are not in fact present. These are called pseudoforces, as described above. Since rotational motion is always accelerated motion, pseudoforces may always be observed in rotating frames of reference.

As one example, a frame of reference in which the Earth is at rest must rotate once per year about the Sun. In this reference frame, the gravitational force attracting the Earth toward the Sun appears to be balanced by an equal and opposite outward force that keeps the Earth in stationary equilibrium. This outward pseudoforce, is the centrifugal force.

The rotation of the Earth about its own axis also causes pseudoforces for observers at rest on the Earth's surface. There is a centrifugal force, but it is much smaller than the force of gravity. Its effect is that, at the Equator, where it is largest, the gravitational acceleration g is about 0.5 percent smaller than at the poles, where there is no centrifugal force. This same centrifugal force is responsible for the fact that the Earth is slightly nonspherical, bulging just a bit at the Equator.

Pseudoforces can have real consequences. The oceanic tides on Earth, for example, are a consequence of centrifugal forces in the Earth-Moon and Earth-Sun systems. The Moon appears to be orbiting the Earth, but in reality both the Moon and the Earth orbit their common centre of mass. The centre of mass of the Earth-Moon system is located inside the Earth nearly three-fourths of the distance from the centre to the surface, or roughly 4,700 kilometres from the centre of the Earth. The Earth rotates about this point approximately once a month. The gravitational attraction of the Moon and the centrifugal force of this rotation are exactly balanced at the centre of the Earth. At the surface of the Earth closest to the Moon, the Moon's gravity is stronger than the centrifugal force. The ocean's waters, which are free to move in response to this unbalanced force, tend to build up a small bulge at that point. On the surface of the Earth exactly opposite the Moon, the centrifugal force is stronger than the Moon's gravity, and a small bulge of water tends to build up there as well. The water is correspondingly depleted at the points 90° on either side of these. Each day the Earth rotates beneath these bulges and troughs, which remain stationary with respect to the Earth-Moon system. The result is two high tides and two low tides every day every place on Earth. The Sun has a similar effect, but of only about half the size; it increases or decreases the size of the tides depending on its relative alignment with the Earth and Moon.

Coriolis Force

The Coriolis force is a pseudoforce that operates in all rotating frames. One way to envision it is to imagine a rotating platform (such as a merry-go-round or a phonograph turntable) with a perfectly smooth surface and a smooth block sliding inertially across it. The block, having no (real) forces acting on it, moves in a straight line at constant speed in inertial space. However, the platform rotates under it, so that to an observer on the platform, the block appears to follow a curved trajectory, bending in the opposite direction to the motion of the platform. Since the motion is curved, and hence accelerated, there appears, to the observer, to be a force operating. That pseudoforce is called the Coriolis force.

The Coriolis force also may be observed on the surface of the Earth. For example, many science museums have a pendulum, called a Foucault pendulum, suspended from a long cable with markers to show that its plane of motion rotates slowly. The rotation of the plane of motion is caused by the Coriolis force. The effect is most easily imagined by picturing the pendulum swinging directly above the North Pole. The plane of its motion remains stationary in inertial space, while the Earth rotates once a day beneath it.

At lower latitudes, the effect is a bit more subtle, but it is still present. Imagine that, somewhere in the Northern Hemisphere, a projectile is fired due south. As viewed from inertial space, the projectile initially has an eastward component of velocity as well as a southward component because the gun that fired it, which is stationary on the surface of the Earth, was moving eastward with the Earth's rotation at the instant it was fired. However, since it was fired to the south, it lands at a slightly lower latitude, closer to the Equator. As one moves south, toward the Equator, the tangential speed of the Earth's surface due to its rotation increases because the surface is farther from the axis of rotation. Thus, although the projectile has an eastward component of velocity (in inertial space), it lands at a place where the surface of the Earth has a larger eastward component of velocity. Thus, to the observer on Earth, the projectile seems to curve slightly to the west. That westward curve is attributed to the Coriolis force. If the projectile were fired to the north, it would seem to curve eastward.

The same analysis applied to a Foucault pendulum explains why its plane of motion tends to rotate in the clockwise direction anywhere in the Northern Hemisphere and in the counterclockwise direction in the Southern Hemisphere. Storms, known as cyclones, tend to rotate in the opposite direction in each hemisphere, also due to the Coriolis force. Air moves in all directions toward a low-pressure centre. In the Northern Hemisphere, air moving up from the south is deflected eastward, while air moving down from the north is deflected westward. This effect tends to give cyclones a counterclockwise circulation in the Northern Hemisphere. In the Southern Hemisphere, cyclones tend to circulate in the clockwise direction.

Spinning Tops and Gyroscopes

Figure below (top) shows a wheel that is weighted in its rim to maximize its moment of inertia I and that is spinning with angular frequency ω on a horizontal axle supported at both ends. As shown, it has an angular momentum L along the x direction equal to Iω. Now suppose the support at point P is removed, leaving the axle supported only at one end. Gravity, acting on the mass of the wheel as if it were concentrated at the centre of mass, applies a downward force on the wheel.

The wheel, however, does not fall. Instead, the axle remains (nearly) horizontal but rotates in the counterclockwise direction as seen from above. This motion is called gyroscopic precession.

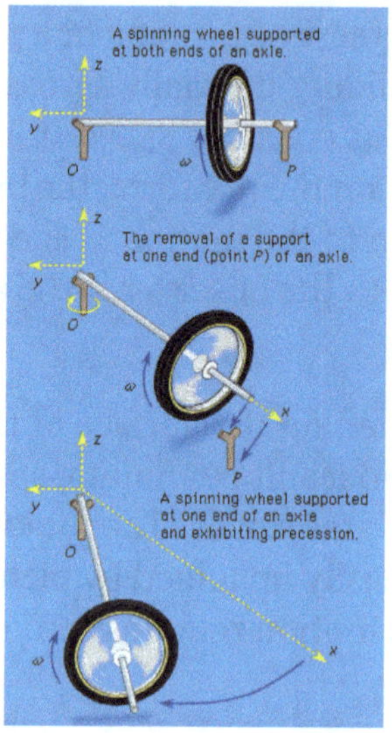

Gyroscopic precession.

Horizontal precession occurs in this case because the gravitational force results in a torque with respect to the point of suspension, such that $\tau = r \times F$ and is directed, initially, in the positive y direction. The torque causes the angular momentum L to move toward that direction according to $\tau = dL/dt$. Because τ is perpendicular to L, it does not change the magnitude of the angular momentum, only its direction. As precession proceeds, the torque remains horizontal, and the angular momentum vector, continually redirected by the torque, executes uniform circular motion in the horizontal plane at a frequency Ω, the frequency of precession.

In reality, the motion is a bit more complicated than uniform precession in the horizontal plane. When the support at P is released, the centre of mass of the wheel initially drops slightly below the horizontal plane. This drop reduces the gravitational potential energy of the system, releasing kinetic energy for the orbital motion of the centre of mass as it precesses. It also provides a small component of L in the negative z direction, which balances the angular momentum in the positive z direction that results from the orbital motion of the centre of mass. There can be no net angular momentum in the vertical direction because there is no component of torque in that direction.

One more complication: the initial drop of the centre of mass carries it too far for a stable plane of precession, and it tends to bounce back up after overshooting. This produces an up-and-down oscillation during precession, called nutation (nodding). In most cases, nutation is quickly damped by friction in the bearings, leaving uniform precession.

A spinning top undergoes all the motions described above. If it is initially set spinning with a vertical axis, there will be virtually no torque, and conservation of angular momentum will keep the axis vertical for a long time. Eventually, however, friction at the point of contact will require the centre of mass to lower itself, which can only happen if the axis tilts. The spinning will also slow down, making the tilting process easier. Once the top tilts, gravity produces a horizontal torque that leads to precession of the spin axis. The subsequent motion depends on whether the point of contact is fixed or free to slip on the horizontal plane. Vast tomes have been written on the motions of tops.

A gyroscope is a device that is designed to resist changes in the direction of its axis of spin. That purpose is generally accomplished by maximizing its moment of inertia about the spin axis and by spinning it at the maximum practical frequency. Each of these considerations has the effect of maximizing the magnitude of the angular momentum, thus requiring a larger torque to change its direction. It is quite generally true that the torque τ, the angular momentum L, and the precession frequency Ω (defined as a vector along the precession axis in the direction given by the right-hand rule) are related by,

$$\tau = \Omega \times L.$$

Equation $\tau = \Omega \times L$, illustrated in figure, is called the gyroscope equation.

$$\tau = \Omega \times L.$$

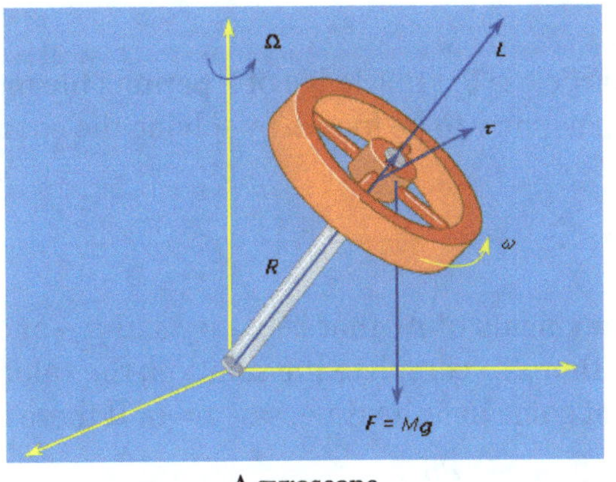

A gyroscope.

Gyroscopes are used for a variety of purposes, including navigation. Use of gyroscopes for this purpose is called inertial guidance. The gyroscope is suspended as nearly as possible at its centre of mass, so that gravity does not apply a torque that causes it to process. The gyroscope tends therefore to point in a constant direction in space, allowing the orientation of the vehicle to be accurately maintained.

One further application of the gyroscope principle may be seen in the precession of the equinoxes. The Earth is a kind of gyroscope, spinning on its axis once each day. The Sun would apply no torque to the Earth if the Earth were perfectly spherical, but it is not. The Earth bulges slightly at the Equator. As indicated in Figure, the effect of the Sun's gravity on the near bulge (larger than it is on the far bulge) results in a net torque about the centre of the Earth. When the Earth is on the other side of the Sun, the net torque remains in the same direction. The torque is small but persistent. It causes the axis of the Earth to process, about one revolution every 25,800 years.

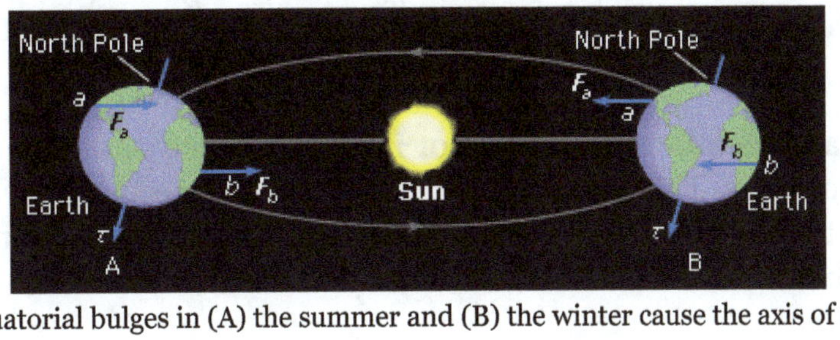

Forces acting on equatorial bulges in (A) the summer and (B) the winter cause the axis of the Earth to precess.

As seen from the Earth, the Sun passes through the plane of the Equator twice each year. These points are called the equinoxes, and on the days of the equinoxes the hours of daylight and night are equal. From antiquity it has been known that the point in the sky where the Sun intersects the plane of the Equator is not the same each year but rather drifts very slowly to the west. This ancient observation, first explained by Newton, is due to the precession of the Earth's axis. It is called the precession of the equinoxes.

Analytic Approaches

Classical mechanics can, in essence, be reduced to Newton's laws, starting with the second law, in the form,

$$F = \frac{dp}{dt}.$$

If the net force acting on a particle is F, knowledge of F permits the momentum p to be found; and knowledge of p permits the position r to be found, by solving the equation,

$$\frac{dr}{dt} = \frac{P}{m}.$$

These solutions give the components of p—that is, p_x, p_y, and p_z—and the components of r—x, y, and z—each as a function of time. To complete the solution, the value of each quantity—p_x, p_y, p_z, x, y, and z—must be known at some definite time, say, $t = 0$. If there is more than one particle, an equation in the form of equation $F = \frac{dp}{dt}$. must be written for each particle, and the solution will involve finding the six variables x, y, z, p_x, p_y, and pz, for each particle as a function of time, each once again subject to some initial condition. The equations may not be independent, however. For example, if the particles interact with one another, the forces will be related by Newton's third law. In this case (and others), the forces may also depend on time.

If the problem involves more than a very few particles, this method of solution quickly becomes intractable. Furthermore, in many cases it is not useful to express the problem purely in terms of particles and forces. Consider, for example, the problem of a sphere or cylinder rolling without slipping on a plane surface. Rolling without slipping is produced by friction due to forces acting between atoms in the rolling body and atoms in the plane, but the interactions are very complex; they probably are not fully understood even today, and one would like to be able to formulate and solve the problem without introducing them or needing to understand them. For all these reasons, methods that go beyond solving equations $F = \frac{dp}{dt}$ and $\frac{dr}{dt} = \frac{P}{m}$ have had to be introduced into classical mechanics.

The methods that have been introduced do not involve new physics. In fact, they are deduced directly from Newton's laws. They do, however, involve new concepts, new language to describe those concepts, and the adoption of powerful mathematical techniques. Some of those methods are briefly surveyed here.

Configuration Space

The position of a single particle is specified by giving its three coordinates, x, y, and z. To specify the positions of two particles, six coordinates are needed, $x_1, y_1, z_1, x_2, y_2, z_2$. If there are N particles, $3N$ coordinates will be needed. Imagine a system of $3N$ mutually orthogonal coordinates in a $3N$-dimensional space (a space of more than three dimensions is a purely mathematical construction, sometimes known as a hyperspace). To specify the exact position of one single point in this space, $3N$ coordinates are needed. However, one single point can represent the entire configuration of all N particles in the problem. Furthermore, the path of that single point as a function of time is the complete solution of the problem. This $3N$-dimensional space is called configuration space.

Configuration space is particularly useful for describing what is known as constraints on a problem. Constraints are generally ways of describing the effects of forces that are best not explicitly introduced into the problem. For example, consider the simple case of a falling body near the surface of the Earth. The equations of motion—equations $z = z_0 - \frac{1}{2}gt^2$, $v = gt$ and $a = g$ —are valid only until the body hits the ground. Physically, this restriction is due to forces between atoms in the falling body and atoms in the ground, but, as a practical matter, it is preferable to say that the solutions are valid only for $z > 0$ (where $z = 0$ is ground level). This constraint, in the form of an inequality, is very difficult to incorporate directly into the equations of the problem. In the language of configuration space, however, one merely needs to specify that the problem is being solved only in the region of configuration space for which $z > 0$.

Notice that the constraint mentioned above, rolling without sliding on a plane, cannot easily be described in configuration space, since it is basically a condition on relative velocities of rotation and translation; but another constraint, that the body is restricted to motion along the plane, is easily described in configuration space.

Another type of constraint specifies that a body is rigid. Then, even though the body is composed of a very large number of atoms, it is not necessary to find separately the x, y, and z coordinate of each atom because these are related to those of the other atoms by the condition of rigidity. A careful analysis yields that, rather than needing $3N$ coordinates (where N may be, for example, 10^{24} atoms), only 6 are needed, 3 to specify the position of the centre of mass and 3 to give the orientation of the body. Thus, in this case, the constraint has reduced the number of independent coordinates from $3N$ to 6. Rather than restricting the behaviour of the system to a portion of the original $3N$-dimensional configuration space, it is possible to describe the system in a much simpler 6-dimensional configuration space. It should be noted, however, that the six coordinates are not necessarily all distances. In fact, the most convenient coordinates are three distances (the x, y, and z coordinates of the centre of mass of the body) and three angles, which specify the orientation of a set of axes fixed in the body relative to a set of axes fixed in space. This is an example of the use of constraints to reduce the number of dynamic variables in a problem (the x, y, and z coordinates of each particle) to a smaller number of generalized dynamic variables, which need not even have the same dimensions as the original ones.

The Principle of Virtual Work

A special class of problems in mechanics involves systems in equilibrium. The problem is to find the configuration of the system, subject to whatever constraints there may be, when all forces are balanced. The body or system will be at rest (in the inertial rest frame of its centre of mass), meaning that it occupies one point in configuration space for all time. The problem is to find that point. One criterion for finding that point, which makes use of the calculus of variations, is called the principle of virtual work.

According to the principle of virtual work, any infinitesimal virtual displacement in configuration space, consistent with the constraints, requires no work. A virtual displacement means an instantaneous change in coordinates (a real displacement would require finite time during which particles might move and forces might change). To express the principle, label the generalized coordinates $r_1, r_2, \ldots, r_i, \ldots$. Then if F_i is the net component of generalized force acting along the coordinate r_i,

$$\sum_i F_i dr_i = 0.$$

Here, $F_i dr_i$ is the work done when the generalized coordinate is changed by the infinitesimal amount dr_i. If r_i is a real coordinate (say, the x coordinate of a particle), then *Fi* is a real force. If r_i is a generalized coordinate (say, an angular displacement of a rigid body), then *Fi* is the generalized force such that $F_i dr_i$ is the work done (for an angular displacement, *Fi* is a component of torque).

Take two simple examples to illustrate the principle. First consider two particles that are restricted to motion in the x direction and are constrained by a taut string connecting them. If their x coordinates are called x_1 and x_2, then $F_1 dx_1 + F_2 dx_2 = 0$ according to the principle of virtual work. But the taut string requires that the particles be displaced the same amount, so that $dx_1 = dx_2$, with the result that $F_1 + F_2 = 0$. The particles might be in equilibrium, for example, under equal and opposite forces, but F_1 and F_2 do not need individually to be zero. This is generally true of the F_i in equation $\sum_i F_i dr_i = 0$. As a second example, consider a rigid body in space. Here, the constraint of rigidity has already been expressed by reducing the coordinate space to that of six generalized coordinates. These six coordinates (x, y, z, and three angles) can change quite independently of one another. In other words, in equation $\sum_i F_i dr_i = 0$, the six *dri* are arbitrary. Thus, the only way equation $\sum_i F_i dr_i = 0$ can be satisfied is if all six F_i are zero. This means that the rigid body can have no net component of force and no net component of torque acting on it. Of course, this same conclusion was reached earlier by less abstract arguments.

Lagrange's and Hamilton's Equations

Elegant and powerful methods have also been devised for solving dynamic problems with constraints. One of the best known is called Lagrange's equations. The Lagrangian L is defined as $L = T - V$, where T is the kinetic energy and V the potential energy of the system in question. Generally speaking, the potential energy of a system depends on the coordinates of all its particles; this may be written as $V = V(x_1, y_1, z_1, x_2, y_2, z_2, \ldots)$. The kinetic energy generally depends on the velocities, which, using the notation $vx = dx/dt = \dot{x}$, may be written $T = T(\dot{x}_1, \dot{y}_1, \dot{z}_1, \dot{x}_2, \dot{y}_2, \dot{z}_2, \ldots)$. Thus, a dynamic problem has six dynamic variables for each particle—that is, x, y, z and $\dot{x}$, $\dot{y}$, $\dot{z}$—and the Lagrangian depends on all $6N$ variables if there are N particles.

In many problems, however, the constraints of the problem permit equations to be written relating at least some of these variables. In these cases, the $6N$ related dynamic variables may be reduced to a smaller number of independent generalized coordinates (written symbolically as $q_1, q_2, \ldots qi, \ldots$) and generalized velocities (written as $\dot{q}_1, \dot{q}_2, \ldots \dot{q}i, \ldots$), just as, for the rigid body, $3N$ coordinates were reduced to six independent generalized coordinates (each of which has an associated velocity). The Lagrangian, then, may be expressed as a function of all the qi and $\dot{q}i$. It is possible, starting from Newton's laws only, to derive Lagrange's equations:

$$\frac{d}{dt}\frac{\partial L}{\partial \dot{q}i} - \frac{\partial L}{\partial qi} = 0.$$

where the notation $\partial L/\partial q_i$ means differentiate L with respect to qi only, holding all other variables constant. There is one equation of the form $\frac{d}{dt}\frac{\partial L}{\partial \dot{q}i} - \frac{\partial L}{\partial qi} = 0$ for each of the generalized coordinates qi (e.g., six equations for a rigid body), and their solutions yield the complete dynamics of the system. The use of generalized coordinates allows many coupled equations of the form $F = \frac{dp}{dt}$ to be reduced to fewer, independent equations of the form $\frac{d}{dt}\frac{\partial L}{\partial \dot{q}i} - \frac{\partial L}{\partial qi} = 0.$

There is an even more powerful method called Hamilton's equations. It begins by defining a generalized momentum pi, which is related to the Lagrangian and the generalized velocity $\dot{q}_i$ by $p_i = \partial L/\partial \dot{q}_i$. A new function, the Hamiltonian, is then defined by $H = \Sigma_i \dot{q}_i p_i - L$. From this point it is not difficult to derive:

$$\dot{q}i = \frac{\partial H}{\partial pi}$$

and

$$-\dot{p}i = \frac{\partial H}{\partial pi}.$$

These are called Hamilton's equations. There are two of them for each generalized coordinate. They may be used in place of Lagrange's equations, with the advantage that only first derivatives not second derivatives are involved.

The Hamiltonian method is particularly important because of its utility in formulating quantum mechanics. However, it is also significant in classical mechanics. If the constraints in the problem do not depend explicitly on time, then it may be shown that H = T + V, where T is the kinetic energy and V is the potential energy of the system—i.e., the Hamiltonian is equal to the total energy of the system. Furthermore, if the problem is isotropic (H does not depend on direction in space) and homogeneous (H does not change with uniform translation in space), then Hamilton's equations immediately yield the laws of conservation of angular momentum and linear momentum, respectively.

Force

Forces acting on objects are vectors that are characterized by not only a magnitude (e.g. Newtons or pounds force) but also a direction. A force vector F (vectors are usually noted by a boldface letter) can be broken down into its components in the x, y and z directions in whatever coordinate system you've drawn:

$$F = F_x i + F_y j + F_z k$$

Where F_x, F_y and F_z are the magnitudes of the forces (units of force, e.g. Newtons or pounds force) in the x, y and z directions and i, j and k are the unit vectors in the x, y and z directions i.e. vectors whose directions are aligned with the x, y and z coordinates and whose magnitudes are exactly 1 (no units).

Forces can also be expressed in terms of the magnitude = $\left(F_x^2 + F_y^2 + F_z^2\right)^{1/2}$ and direction relative to the positive x-axis (= $\tan^{-1}\left(F_y / F_x\right)$ in a 2-dimensional system). Note that the $\tan^{-1}\left(F_y / F_x\right)$ function gives you an angle between +90° and -90° whereas sometimes the resulting force is between +90° and +180° or between -90° and -180°; in these cases you'll have to examine the resulting force and add or subtract 180° from the force to get the right direction.

Degrees of Freedom

Imagine a one-dimensional (1D) world, i.e. where objects can move (translate) back and forth along a single line but in no other way. For this 1D world there is only one direction (call it the xdirection) that the object can move linearly and no way in which it can rotate, hence only one force balance equation is required. For the field of dynamics this equation would be Newton's Second Law, namely that the sum of the forces $F_{x,1} + F_{x,2} + F_{x,3} + \ldots + F_{x,n}$ = max where m is the mass of the object and ax is the acceleration of the object in the x direction.

So a 1D world is quite simple, but what about a 2D world? Do we just need a second force balance equation for translation in the y direction (that is, SF_y = 0) and we're done? Well, no. Let's look at a counter-example. The set of forces on the object in the left panel satisfies the requirements $\Sigma F_Y = 0$ and $\Sigma F_Y = 0$ and would appear to be in static equilibrium. In the right panel, it is also true that $\Sigma F_Y = 0$ and $\Sigma F_Y = 0$, but clearly this object would not be stationary; instead it would be rotating clockwise. Why is this? In two dimensions, in addition to the translational degrees of freedom in the x and y directions, there is also one rotational degree of freedom, that is, the object can rotate about an axis perpendicular to the x-y plane, i.e., an axis in the z-direction. How do we ensure that the object is not rotating? We need to account for the moments of force (M) in addition to the forces themselves, and just as the forces in the x and y directions must add up to zero, i.e. $\Sigma F_Y = 0$ and $\Sigma F_Y = 0$, we need to have the moments of force add up to zero, i.e. $\Sigma M = 0$. So to summarize, in order to have static equilibrium of an object, the sum of all the forces and the moments of force must be zero. In other words, there are two ways that a 2-dimensional object can translate (in the x and y directions) and one way that in can rotate (with the axis of rotation perpendicular to the x-y plane.) So there are 3 equations that must be satisfied in order to have equilibrium,

$$\sum_{i=1}^{n} F_{x,i} = 0; \sum_{i=1}^{n} F_{y,i} = 0; \sum_{i=1}^{n} M_{i,A} = 0$$

where the number of forces in the x direction is n, the number of forces in the y direction is m and p = n + m is the number of moments of force calculated with respect to some point A in the (x,y) plane. (The choice of location of point A is discussed below, but the bottom line is that any point yields the same result.

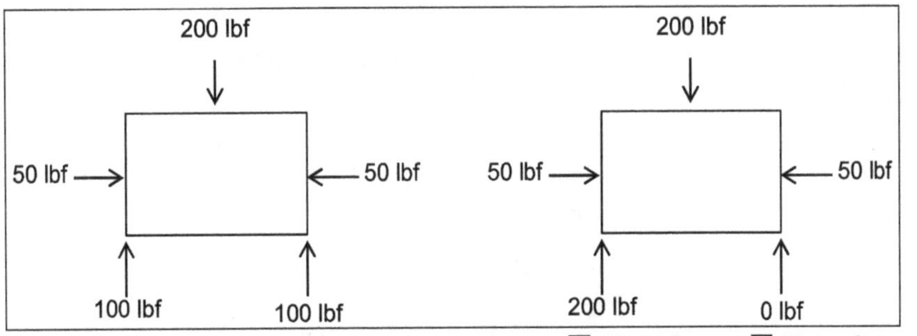

Two sets of forces on an object, both satisfying $\sum F_Y = 0$ and $\sum F_Y = 0$, but one (left) in static equilibrium, the other (right) not in static equilibrium.

This is all fine and well for a 2D (planar) situation, what about 3D? For 3D, there are 3 directions an object can move linearly (translate) and 3 axes about which it can rotate, thus we need 3 force balance equations (in the x, y and z directions) and 3 moment of force balance equations (one each about the x, y and z axes.) Table below summarizes these situations.

Numbers of spatial dimensions	Maximum numbers of force balances	Minimum Numbers of moment of force balances	Total Numbers of unknown forces & moments
1	1	0	1
2	2	1	3
3	3	3	6

Table: Number of force and moment of force balance equations required for static equilibrium as a function of the dimensionality of the system. (But note that, as just described, moment of force balance equations can be substituted for force balance equations).

Moments of Forces

Some types of structures can only exert forces along the line connecting the two ends of the structure, but cannot exert any force perpendicular to that line. These types of structures include ropes, ends with pins, and bearings. Other structural elements can also exert a force perpendicular to the line. This is called the moment of force (often shortened to just "moment", but to avoid confusion with "moment" meaning a short period of time, we will use the full term "moment of force") which is the same thing as torque. Usually the term torque is reserved for the forces on rotating, not stationary, shafts, but there is no real difference between a moment of force and a torque.

The distinguishing feature of the moment of force is that it depends not only on the vector force itself (F_i) but also the distance (d_i) from that line of force to a reference point A. from the anchor point at which it acts. If you want to loosen a stuck bolt, you want to apply whatever force your arm is capable of providing over the longest possible di. The line through the force F_i is called the line of action. The moment arm is the distance (d_i again) between the line of action and a line parallel to the line of action that passes through the anchor point. Then the moment of force (M_i) is defined as:

$$M_i = F_i d_i$$

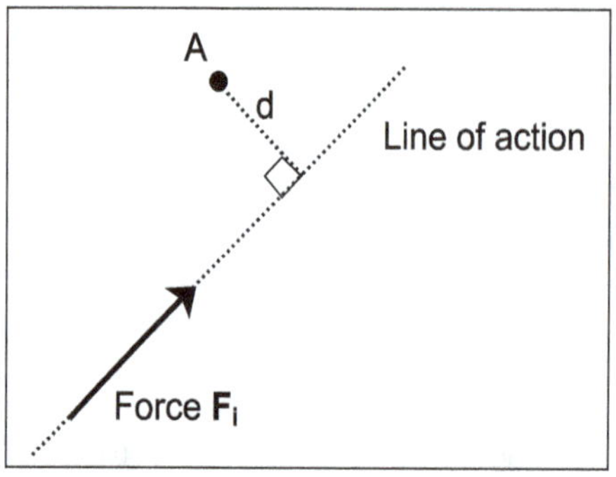

Force, line of action and moment of force (= Fd) about a point A. The example shown is a counterclockwise moment of force, i.e. force Fi is trying to rotate the line segment d counterclockwise about point A.

where F_i is the magnitude of the vector F. Note that the units of Mo is force x length, e.g. ft lbf or N m. This is the same as the unit of energy, but the two have nothing in common – it's just coincidence. So one could report a moment of force in units of Joules, but this is unacceptable practice – use N m, not J.

Note that it is necessary to assign a sign to M_i depending on whether the moment of force is trying to rotate the free body clockwise or counter-clockwise. Typically we will define a clockwise moment of force as positive and counterclockwise as negative, but one is free to choose the opposite definition – as long as you're consistent within an analysis.

Note that the moment of forces must be zero regardless of the choice of the origin (i.e. not just at the center of mass). So one can take the origin to be wherever it is convenient (e.g. make the moment of one of the forces = 0.) Consider the very simple set of forces below:

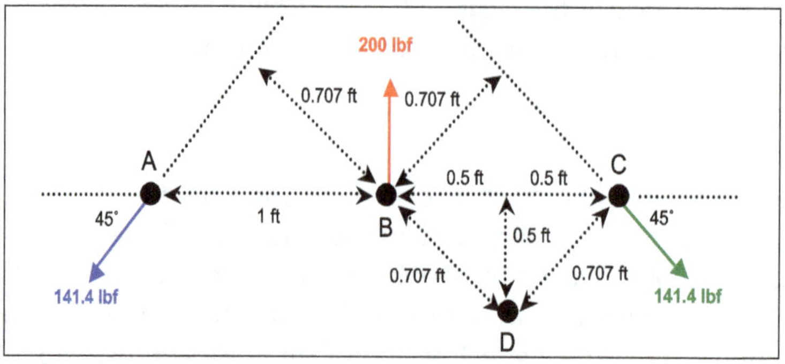

Force diagram showing different ways of computing moments of force.

Because of the symmetry, it is easy to see that this set of forces constitutes an equilibrium condition. When taking moments of force about point 'B' we have:

$$\sum F_x = +141.4\cos(45^\circ)\ \text{lbf} + 0 - 141.4\cos(45^\circ)\ \text{lbf} = 0$$
$$\sum F_y = +141.4\sin(45^\circ)\ \text{lbf} - 200\ \text{lbf} + 141.4\sin(45^\circ) = 0$$
$$\sum M_B = -141.4\ \text{lbf} * 0.707\ \text{ft} - 200\ \text{lbf} * 0\ \text{ft} + 141.4\ \text{lbf} * 0.707\ \text{ft} = 0.$$

But how do we know to take the moments of force about point B? We don't. But notice that if we take the moments of force about point 'A' then the force balances remain the same and

$$\sum M_A = -141.4\ \text{lbf} * 0\ \text{ft} - 200\ \text{lbf} * 1\ \text{ft} + 141.4\ \text{lbf} * 1.414\ \text{ft} = 0.$$

The same applies if we take moments of force about point 'C', or a point along the line ABC, or even a point NOT along the line ABC. For example, taking moments of force about point 'D',

$$\sum M_D = -141.4\ \text{lbf} * (0.707\ \text{ft} + 0.707\ \text{ft}) + 200\ \text{lbf} * 0.5\ \text{ft} + 141.4\ \text{lbf} * 0.707\ \text{ft} = 0$$

The location about which to take the moments of force can be chosen to make the problem as simple as possible, e.g. to make some of the moments of forces = 0.

Example of "why didn't the book just say that?" The state of equilibrium merely requires that 3 constraint equations are required. There is nothing in particular that requires there must be 2 force and 1 moment of force constraint equations. So one could have 1 force and 2 moment of force constraint equations:

$$\sum_{i=1}^{n} F_{x,i} = 0; \sum_{i=1}^{n} M_{i,A} = 0; \sum_{i=1}^{n} M_{j,B} = 0$$

where the coordinate direction x can be chosen to be in any direction, and moments of force are taken about 2 separate points A and B. Or one could even have 3 moment of force equations:

$$\sum_{i=1}^{n} F_{i,A} = 0; \sum_{j=1}^{n} M_{j,B} = 0; \sum_{j=1}^{n} M_{j,C} = 0.$$

Also, there is no reason to restrict the x and y coordinates to the horizontal and vertical directions. They can be (for example) parallel and perpendicular to an inclined surface if that appears in the problem. In fact, the x and y axes don't even have to be perpendicular to each other, as long as they are not parallel to each other, in which case $\sum F_x = 0$ and $\sum F_y = 0$ would not be independent equations.

Types of Forces and Moments of Force

A free body diagram is a diagram showing all the forces and moments of forces acting on an object. We distinguish between two types of objects:

1. Particles that have no spatial extent and thus have no moment arm (d). An example of this

would be a satellite orbiting the earth because the spatial extent of the satellite is very small compared to the distance from the earth to the satellite or the radius of the earth. Particles do not have moments of forces and thus do not rotate in response to a force.

2. Rigid bodies that have a finite dimension and thus has a moment arm (d) associated with each applied force. Rigid bodies have moments of forces and thus can rotate in response to a force.

There are several types of forces that act on particles or rigid bodies:

1. Rope, cable, etc. – Force (tension) must be a long line of action; no moment of force (1 unknown force)

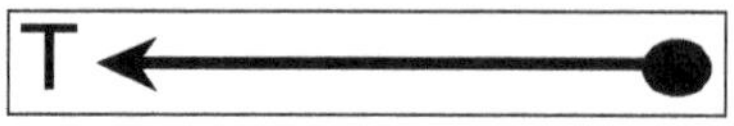

2. Rollers, frictionless surface – Force must be perpendicular to the surface; no moment of force (1 unknown force). There cannot be a force parallel to the surface because the roller would start rolling! Also the force must be away from the surface towards the roller (in other words the roller must exert a force on the surface), otherwise the roller would lift off of the surface.

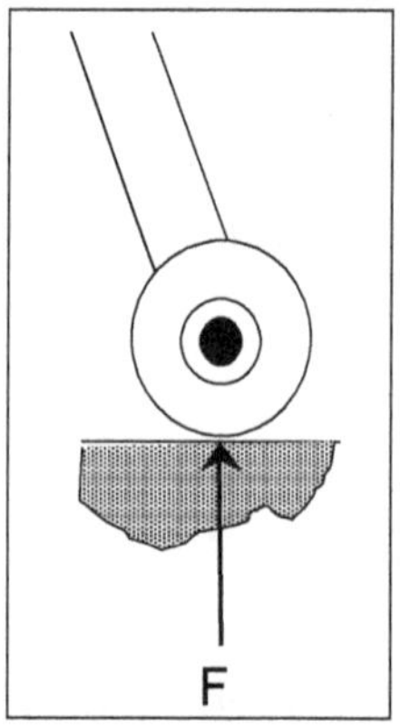

3. Frictionless pin or hinge – Force has components both parallel and perpendicular to the line of action; no moment of force (2 unknown forces) (note that the coordinate system does not need to be parallel and perpendicular to either the gravity vector or the bar)

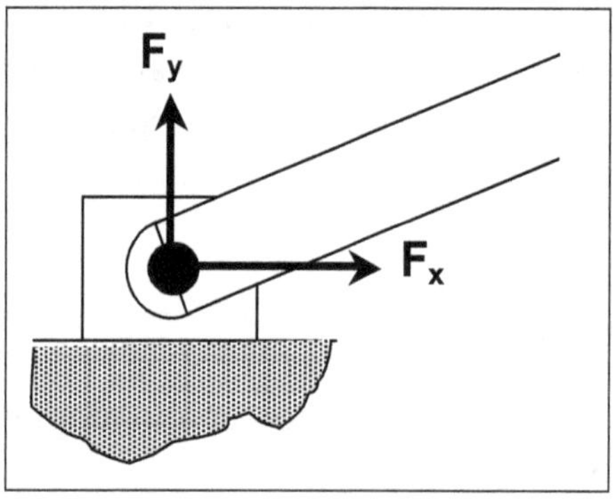

4. Fixed support – Force has components both parallel and perpendicular to line of action

plus a moment of force (2 unknown forces, 1 unknown moment force). Note that for our simple statics problems with 3 degrees of freedom, if there is one fixed support then we already have 3 unknown quantities and the rest of our free body cannot have any unknown forces if we are to employ statics alone to determine the forces. In other words, if the free body has any additional unknown forces the system is statically indeterminate as will be discussed shortly.

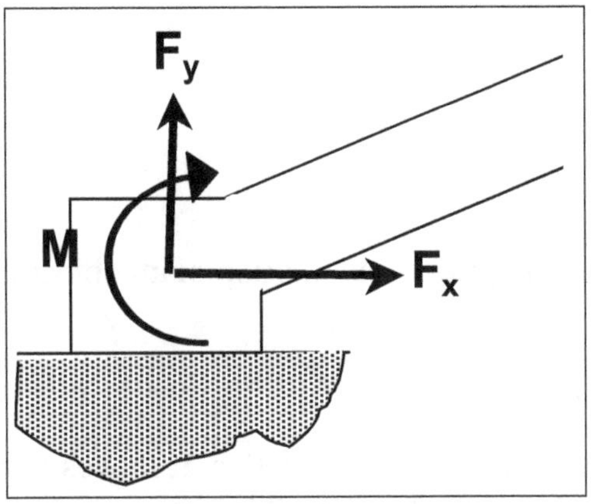

5. Contact friction – Force has components both parallel (F) and perpendicular (N) to surface, which are related by F = μN, where μ is the coefficient of friction, which is usually assigned separate values for static (no sliding) (μ^s) and dynamic (sliding) (μ_d) friction, with the latter being lower. (2 unknown forces coupled by the relation F = μN). μ depends on both of the surfaces in contact. Most dry materials have friction coefficients between 0.3 and 0.6 but Teflon in contact with Teflon, for example, can have a coefficient as low as 0.04. Rubber (e.g. tires) in contact with other surfaces (e.g. asphalt) can yield friction coefficients of almost 2.

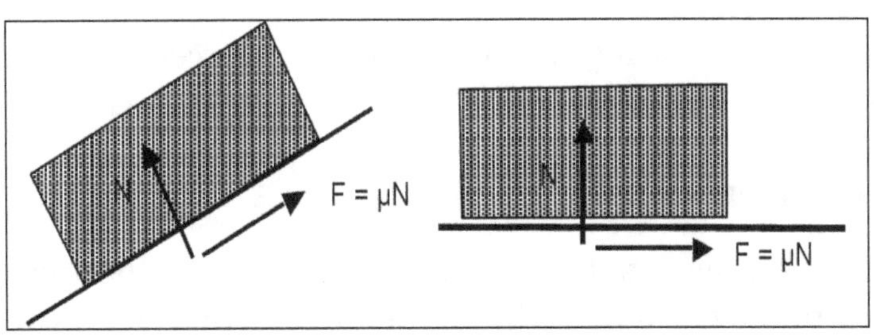

Actually the statement $F = \mu_s N$ for static friction is not correct at all, although that's how it's almost always written. Consider the figure on the right, above. If there is no applied force in the horizontal direction, there is no need for friction to counter that force and keep the block from sliding, so F = 0. (If F ≠ 0, then the object would start moving even though there is no applied force!) Of course, if a force were applied (e.g. from right to left, in the –x direction) then the friction force at the interface between the block and the surface would counter the applied force with a force in the + x direction so that $\sum F_x = 0$. On the other hand, if a force were applied from left to right, in the + x direction) then the friction force at the interface between the block and the surface would counter the applied force with a force in the -x direction so that $\sum F_x = 0$. The expression $F = \mu_s N$ only applies to the maximum magnitude of the static friction force. In other words, a proper statement quantifying the friction force would be $|F| \leq \mu_s N$, not $F = \mu_s N$. If any larger force is applied then the block would start moving and then the dynamic friction force F = $\mu_d N$ is the applicable one – but even then this force must always be in the direction opposite the motion – so $|F| = \mu_d N$ is an appropriate statement. Another, more precise way of writing this would be $\vec{F} = \mu_d N \frac{\vec{v}}{v}$, where $\vec{v}$ is the velocity of the block and v is the magnitude of this velocity, thus $\frac{\vec{v}}{v}$ is a unit vector in the direction of motion.

Friction

A force of friction is any force that opposes the motion of an object due to the contact of the object with other bodies.

These are the following experiments:

1) Push an empty box on a smooth floor and then let go. It moves 1 or 2 meters and then stops. The force that stops the box is called force of friction. It is a force that opposes any motion.

2) Put some books (or any objects that have some weight) inside the box and push it again then let go. The box will move a distance smaller than when it was empty. The force of friction has acted again to stop the box. This time the force of friction is larger because it is proportional to the normal force exerted by the floor on the box which itself is equal to the weight of the box.

3) Try to push a large double bed, for example, and its does not move. In this case the force of friction is much larger than the force used to push the bed and therefore we talk here about static force of friction exerted by the floor on the bed.

Depending on whether object is moving or static, two types of forces of friction are possible: kinetic force of friction and static force of friction.

Kinetic Force of Friction

The force applied to the box is enough to move the box. In this case the applied force is larger that the force of friction and the object moves. The force of friction that acts on a moving object to slow it down is called kinetic friction force and is given by the formula:

$$F = \mu k\, N,$$

where μ_k is the coefficient of kinetic friction and N is the force normal to the surface acting on the box.

Example:

1. Draw a free body diagram of a box of mass M Kg resting on the floor and draw and label all forces acting on it.
2. A force F_a horizontal to the floor is now applied to push the box from right to left. Force F_a is larger that the force of friction. The floor has a coefficient of kinetic friction μ_k.

A free body diagram and all forces acting on the box.

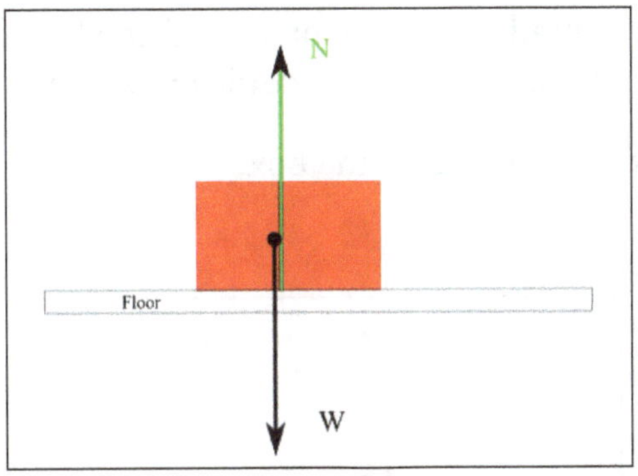

The red rectangle represents the box.

W = weight of the box,

N is the normal exerted by the floor on the box, otherwise the box falls through the floor.

We now present a diagram with the applied force to push the box and the force of kinetic force of friction opposing the motion of the box.

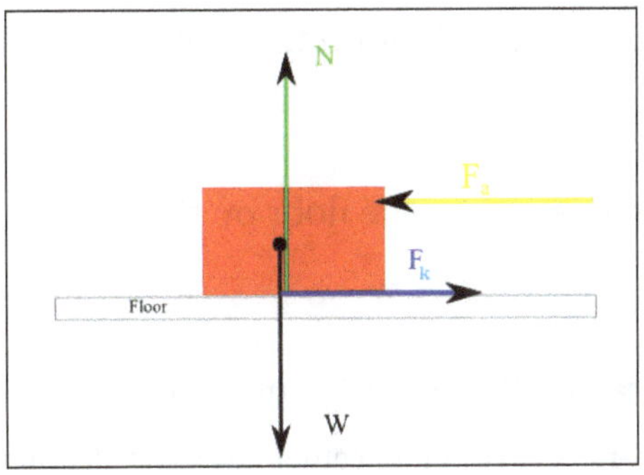

W and N are the weight and the normal force respectively,

F_a the force applied to push the box,

F_k is the force of friction exerted by the floor on the box opposite the direction of motion and is given by,

$$F_k = \mu_k N.$$

Static Force of Friction

A small force F_a is applied to pull a box but this force is not enough to move the box. The force of friction exists and it acts opposite F_a and the box does not move. Because the box does not move, the force of friction in this case is called Static Force of Friction and is given by,

$$F = \mu_s N$$

where μ_s is the coefficient of static friction and N is the force normal to the surface acting on the box.

Example:

A force F_a is now applied horizontally to the floor to pull the box from left to right . Force F_a is smaller than the force of friction. The floor has a coefficient of static friction μ_s.

Draw a diagram and label all forces acting on the box.

Solution:

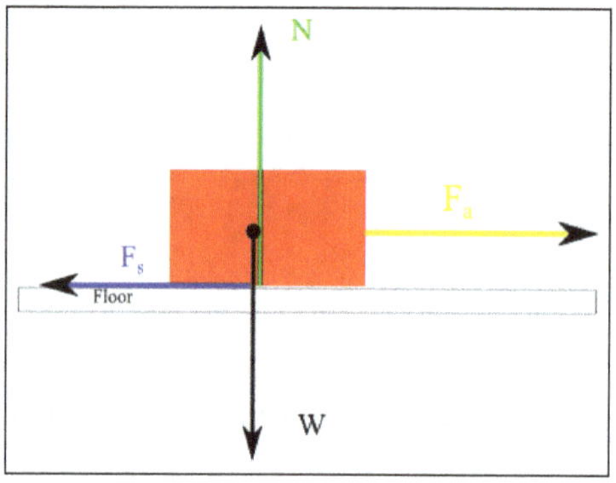

W the weight of the box.

N the force normal to and exerted by the floor on the box.

F_a the applied force to pull the box.

$F_s = \mu_s N$; static force of friction exerted by the floor on the box in the direction opposite to F_a.

References

- Mechanics, science: britannica.com, Retrieved 18 April, 2019
- Motion, physics: physics-and-radio-electronics.com, Retrieved 7 February, 2019
- Motion-of-a-particle-in-one-dimension, mechanics, science: britannica.com, Retrieved 14 January, 2019
- Rigid-bodies, mechanics, science: britannica.com, Retrieved 4 June, 2019
- Force-friction, forces: problemsphysics.com, Retrieved 20 March, 2019

3

Strength of Materials

The ability of withstanding an applied load without deformation or failure is referred to as strength of material. Some of the key areas of study related to the strength of materials are stress, strain and stress-strain curve. This chapter closely examines these important concepts related to strength of materials to provide an extensive understanding of the subject.

Strength of Materials, also referred to as Mechanics of Materials, looks at the behavior of materials when forces are applied to them. These forces include different types of stresses upon and within the material in all different directions, and the strain that is experienced in the material due to those forces. Studying the reactions of a material usually begins with looking at the static forces on and within the material to determine all of the forces affecting it. Once this examination is complete, engineers can find the reactions of the material.

Stress

Stress is the internal resistance offered by the body to the external load applied to it per unit cross sectional area. Stresses are normal to the plane to which they act and are tensile or compressive in nature.

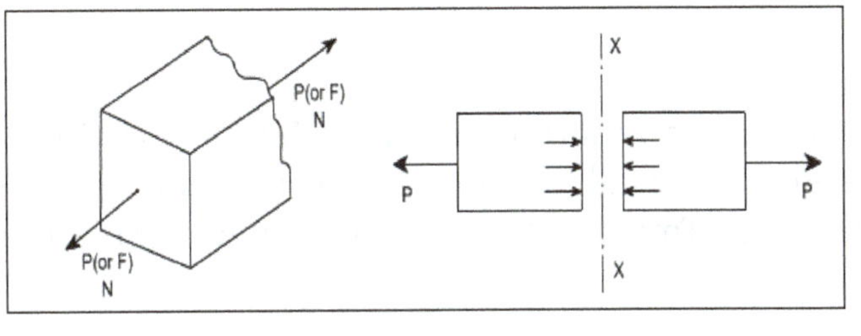

As we know that in mechanics of deformable solids, externally applied forces acts on a body and body suffers a deformation. From equilibrium point of view, this action should be opposed or reacted by internal forces which are set up within the particles of material due to cohesion. These internal forces give rise to a concept of stress. Consider a rectangular rod subjected to axial pull P. Let us imagine that the same rectangular bar is assumed to be cut into two halves at section XX. The each portion of this rectangular bar is in equilibrium under the action of load P and the internal forces acting at the section XX has been shown.

Now stress is defined as the force intensity or force per unit area. Here we use a symbol σ to represent the stress,

$$\sigma = \frac{P}{A}$$

where A is the area of the X –X section.

Here we are using an assumption that the total force or total load carried by the rectangular bar is uniformly distributed over its cross – section. But the stress distributions may be for from uniform, with local regions of high stress known as stress concentrations. If the force carried by a component is not uniformly distributed over its cross – sectional area, A, we must consider a small area, 'δA' which carries a small load 'δP', of the total force 'P', Then definition of stress is,

$$\sigma = \frac{\delta F}{\delta A}$$

As a particular stress generally holds true only at a point, therefore it is defined mathematically as,

$$\sigma = \lim_{\delta A \to 0} \frac{\delta F}{\delta A}$$

Units:

The basic units of stress in S.I units i.e. (International system) are N / m² (or Pa)

MPa = 10^6 Pa

GPa = 10^9 Pa

KPa = 10^3 Pa

Sometimes N / mm² units are also used, because this is an equivalent to MPa. While US customary unit is pound per square inch psi.

Types of Stresses

Only two basic stresses exists: (1) normal stress and (2) shear stress. Other stresses either are similar to these basic stresses or are a combination of this e.g. bending stress is a combination tensile, compressive and shear stresses. Torsional stress, as encountered in twisting of a shaft is a shearing stress.

1. Normal stresses: We have defined stress as force per unit area. If the stresses are normal to the areas concerned, then these are termed as normal stresses. The normal stresses are generally denoted by a Greek letter (σ).

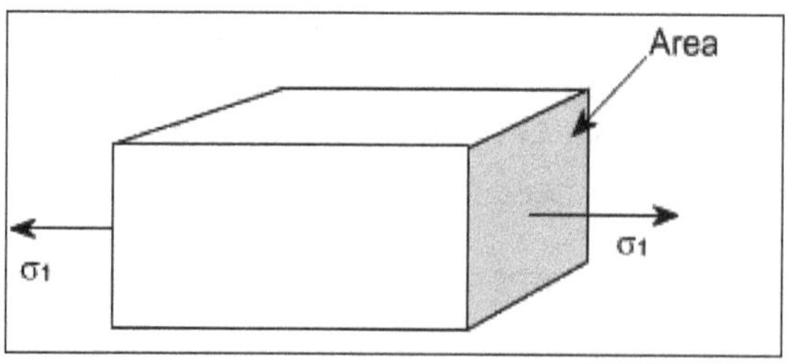

This is also known as uniaxial state of stress, because the stresses acts only in one direction however, such a state rarely exists, therefore we have biaxial and triaxial state of stresses where either the two mutually perpendicular normal stresses acts or three mutually perpendicular normal stresses acts as shown in the figures below :

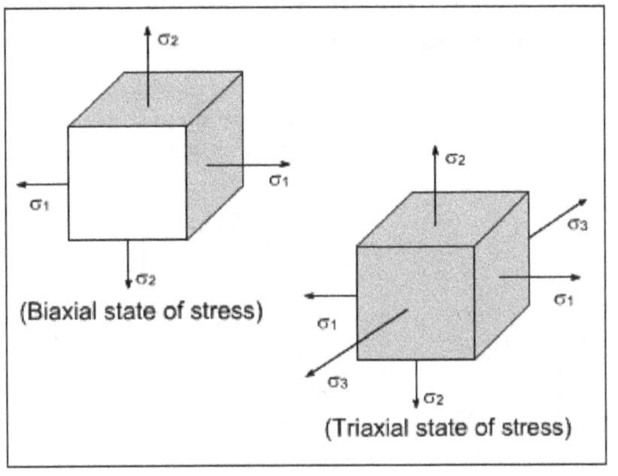

(Biaxial state of stress) (Triaxial state of stress)

2. Tensile or compressive Stresses: The normal stresses can be either tensile or compressive whether the stresses acts out of the area or into the area.

Sign convections for Normal stress:Direct stresses or normal stresses:

- tensile +ve.
- compressive –ve.

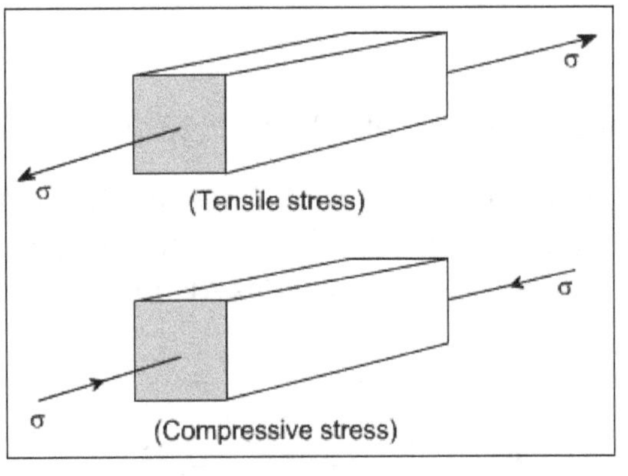

(Tensile stress)

(Compressive stress)

3. Bearing Stress: When one object presses against another, it is referred to a bearing stress (They are in fact the compressive stresses).

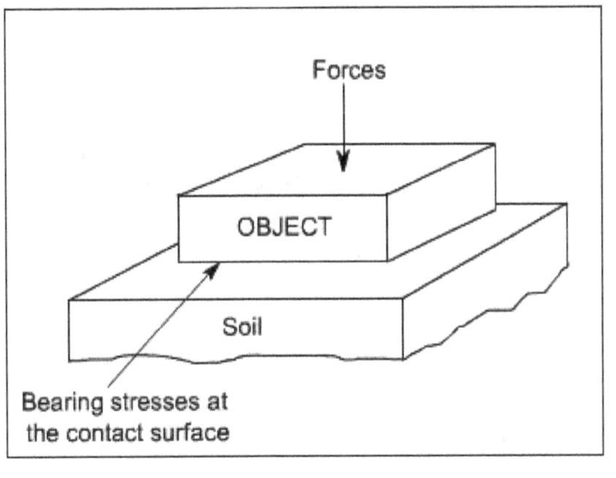

4. Shear Stresses: Let us consider now the situation, where the cross – sectional area of a block of material is subject to a distribution of forces which are parallel, rather than normal, to the area concerned. Such forces are associated with a shearing of the material, and are referred to as shear forces. The resulting stress is known as shear stress.

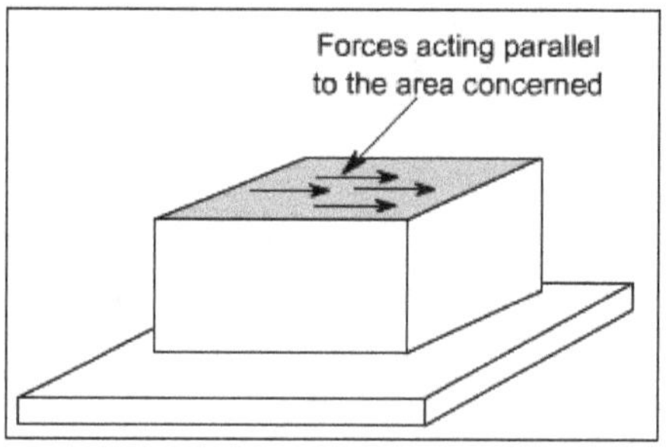

The resulting force intensities are known as shear stresses, the mean shear stress being equal to,

$$\tau = \frac{P}{A}$$

where P is the total force and A the area over which it acts. As we know that the particular stress generally holds good only at a point therefore we can define shear stress at a point as,

$$\tau = \lim_{\delta A \to 0} \frac{\delta F}{\delta A}$$

The Greek symbol τ (tau, suggesting tangential) is used to denote shear stress.

5. Complementary shear stresses: The existence of shear stresses on any two sides of the element induces complementary shear stresses on the other two sides of the element to maintain equilibrium. As shown in the figure the shear stress τ in sides AB and CD induces a complimentary shear stress ' τ in sides AD and BC.

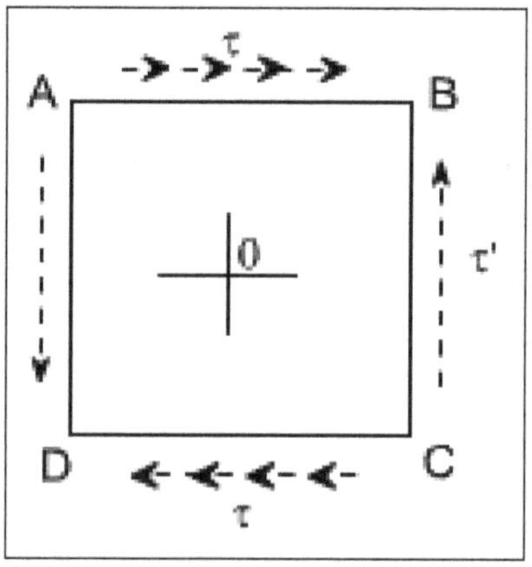

Sign convections for shear stresses:

- Tending to turn the element C.W +ve.
- Tending to turn the element C.C.W – ve.

Deformation of a Body due to Self Weight

Consider a bar AB hanging freely under its own weight as shown in the figure.

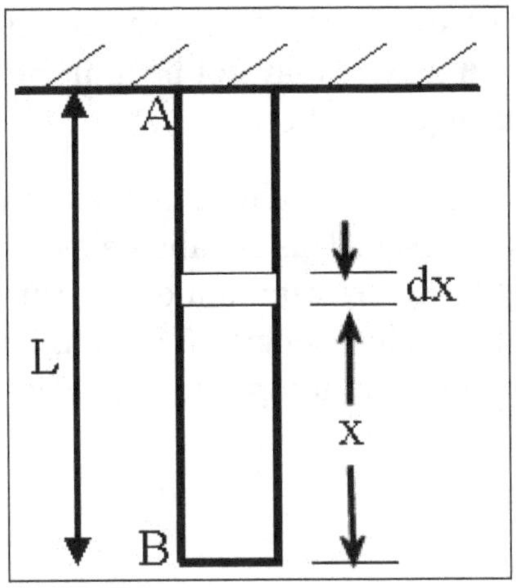

Let,

L= length of the bar.

A= cross-sectional area of the bar.

E = Young's modulus of the bar material.

w= specific weight of the bar material.

Then deformation due to the self-weight of the bar is,

$$\delta L = \frac{WL}{2E}.$$

Strain

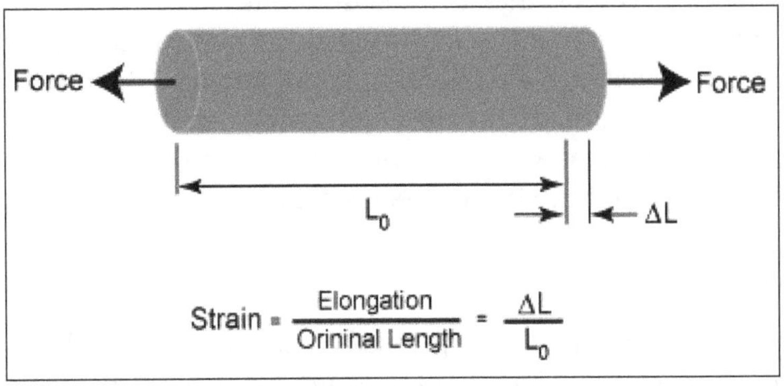

Strain is the response of a system to an applied stress. When a material is loaded with a force, it produces a stress, which then causes a material to deform. Engineering strain is defined as the amount of deformation in the direction of the applied force divided by the initial length of the material. This results in a unitless number, although it is often left in the unsimplified form, such as inches per inch or meters per meter. For example, the strain in a bar that is being stretched in tension is the amount of elongation or change in length divided by its original length. As in the case of stress, the strain distribution may or may not be uniform in a complex structural element, depending on the nature of the loading condition.

If the stress is small, the material may only strain a small amount and the material will return to its original size after the stress is released. This is called elastic deformation, because like elastic it returns to its unstressed state. Elastic deformation only occurs in a material when stresses are lower than a critical stress called the yield strength. If a material is loaded beyond it elastic limit, the material will remain in a deformed condition after the load is removed. This is called plastic deformation.

Strain Measures

Depending on the amount of strain, or local deformation, the analysis of deformation is subdivided into three deformation theories:

- Finite strain theory, also called large strain theory, large deformation theory, deals with deformations in which both rotations and strains are arbitrarily large. In this case, the undeformed and deformed configurations of the continuum are significantly different and a clear distinction has to be made between them. This is commonly the case with elastomers, plastically-deforming materials and other fluids and biological soft tissue.
- Infinitesimal strain theory, also called small strain theory, small deformation theory, small displacement theory, or small displacement-gradient theory where strains and rotations are both small. In this case, the undeformed and deformed configurations of the body can be assumed identical. The infinitesimal strain theory is used in the analysis of deformations of materials exhibiting elastic behavior, such as materials found in mechanical and civil engineering applications, e.g. concrete and steel.
- Large-displacement or large-rotation theory, which assumes small strains but large rotations and displacements.

In each of these theories the strain is then defined differently. The engineering strain is the most common definition applied to materials used in mechanical and structural engineering, which are subjected to very small deformations. On the other hand, for some materials, e.g. elastomers and polymers, subjected to large deformations, the engineering definition of strain is not applicable, e.g. typical engineering strains greater than 1%, thus other more complex definitions of strain are required, such as stretch, logarithmic strain, Green strain, and Almansi strain.

Engineering Strain

The Cauchy strain or engineering strain is expressed as the ratio of total deformation to the initial dimension of the material body in which the forces are being applied. The *engineering normal strain* or *engineering extensional strain* or *nominal strain* e of a material line element or fiber axially loaded is expressed as the change in length ΔL per unit of the original length L of the line element or fibers. The normal strain is positive if the material fibers are stretched and negative if they are compressed. Thus, we have:

$$e = \frac{\Delta L}{L} = \frac{l - L}{L}$$

where e is the engineering normal strain, L is the original length of the fiber and l is the final length of the fiber. Measures of strain are often expressed in parts per million or microstrains.

The true shear strain is defined as the change in the angle (in radians) between two material line elements initially perpendicular to each other in the undeformed or initial configuration. The engineering shear strain is defined as the tangent of that angle, and is equal to the length of deformation at its maximum divided by the perpendicular length in the plane of force application which sometimes makes it easier to calculate.

Stretch Ratio

The stretch ratio or extension ratio is a measure of the extensional or normal strain of a differential line element, which can be defined at either the undeformed configuration or the deformed configuration. It is defined as the ratio between the final length *l* and the initial length *L* of the material line:

$$\lambda = \frac{l}{L}$$

The extension ratio is approximately related to the engineering strain by,

$$e = \frac{l - L}{L} = \lambda - 1$$

This equation implies that the normal strain is zero, so that there is no deformation when the stretch is equal to unity.

The stretch ratio is used in the analysis of materials that exhibit large deformations, such as elastomers, which can sustain stretch ratios of 3 or 4 before they fail. On the other hand, traditional engineering materials, such as concrete or steel, fail at much lower stretch ratios.

True Strain

The logarithmic strain ε, also called, *true strain* or *Hencky strain* . Considering an incremental strain (Ludwik),

$$\delta\varepsilon = \frac{\delta l}{l}$$

the logarithmic strain is obtained by integrating this incremental strain:

$$\int \delta\varepsilon = \int_L^l \frac{\delta l}{l}$$

$$\varepsilon = \ln\left(\frac{l}{L}\right) = \ln(\lambda)$$
$$= \ln(1+e)$$
$$= e - \frac{e^2}{2} + \frac{e^3}{3} - \cdots$$

where *e* is the engineering strain. The logarithmic strain provides the correct measure of the final strain when deformation takes place in a series of increments, taking into account the influence of the strain path.

Green Strain

The Green strain is defined as:

$$\varepsilon \;= \tfrac{1}{2}\left(\frac{l^2 \quad L^2}{}\right) = \tfrac{1}{2}(\lambda \;\; -1).$$

Almansi Strain

The Euler-Almansi strain is defined as,

$$\varepsilon_E = \tfrac{1}{2}\left(\frac{l^2 - L^2}{l^2}\right) = \tfrac{1}{2}\left(1 - \frac{1}{\lambda^2}\right).$$

Normal and Shear Strain

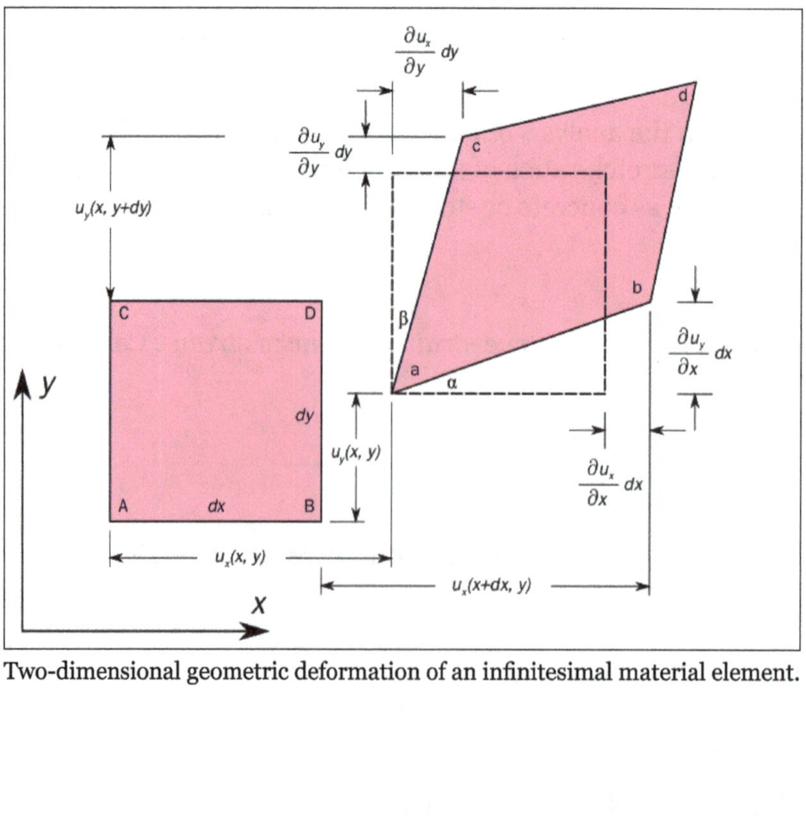

Two-dimensional geometric deformation of an infinitesimal material element.

Strains are classified as either *normal* or *shear*. A *normal strain* is perpendicular to the face of an element, and a *shear strain* is parallel to it. These definitions are consistent with those of normal stress and shear stress.

Normal Strain

For an isotropic material that obeys Hooke's law, a normal stress will cause a normal strain. Normal strains produce *dilations*.

Consider a two-dimensional, infinitesimal, rectangular material element with dimensions $dx \times dy$, which, after deformation, takes the form of a rhombus. The deformation is described by the displacement field u. From the geometry of the adjacent figure we have,

$$\text{length}(AB) = dx$$

and

$$\text{length}(ab) = \sqrt{\left(dx + \frac{\partial u_x}{\partial x} dx\right)^2 + \left(\frac{\partial u_y}{\partial x} dx\right)^2}$$

$$= dx\sqrt{1 + 2\frac{\partial u_x}{\partial x} + \left(\frac{\partial u_x}{\partial x}\right)^2 + \left(\frac{\partial u_y}{\partial x}\right)^2}$$

For very small displacement gradients the squares of the derivatives are negligible and we have,

$$\text{length}(ab) \approx dx + \frac{\partial u_x}{\partial x} dx$$

The normal strain in the x-direction of the rectangular element is defined by,

$$\varepsilon_x = \frac{\text{extension}}{\text{original length}} = \frac{\text{length}(ab) - \text{length}(AB)}{\text{length}(AB)} = \frac{\partial u_x}{\partial x}$$

Similarly, the normal strain in the y- and z-directions becomes,

$$\varepsilon_y = \frac{\partial u_y}{\partial y} \quad , \quad \varepsilon_z = \frac{\partial u_z}{\partial z}.$$

Shear Strain

The engineering shear strain (γ_{xy}) is defined as the change in angle between lines $\overline{AC}$ and $\overline{AB}$. Therefore,

$$\gamma_{xy} = \alpha + \beta$$

From the geometry of the figure, we have,

$$\tan\alpha = \frac{\frac{\partial u_y}{\partial x}dx}{dx + \frac{\partial u_x}{\partial x}dx} = \frac{\frac{\partial u_y}{\partial x}}{1 + \frac{\partial u_x}{\partial x}}$$

$$\tan\beta = \frac{\frac{\partial u_x}{\partial y}dy}{dy + \frac{\partial u_y}{\partial y}dy} = \frac{\frac{\partial u_x}{\partial y}}{1 + \frac{\partial u_y}{\partial y}}$$

For small displacement gradients we have,

$$\frac{\partial u_x}{\partial x} \ll 1 \; ; \; \frac{\partial u_y}{\partial y} \ll 1$$

For small rotations, i.e. α and β are $\ll 1$ we have $\tan\alpha \approx \alpha$, $\tan\beta \approx \beta$. Therefore,

$$\alpha \approx \frac{\partial u_y}{\partial x} \; ; \; \beta \approx \frac{\partial u_x}{\partial y}$$

Thus,

$$\gamma_{xy} = \alpha + \beta = \frac{\partial u_y}{\partial x} + \frac{\partial u_x}{\partial y}$$

By interchanging x and y and u_x and u_y, it can be shown that $\gamma_{xy} = \gamma_{yx}$.

Similarly, for the yz- and xz-planes, we have:

$$\gamma_{yz} = \gamma_{zy} = \frac{\partial u_y}{\partial z} + \frac{\partial u_z}{\partial y} \quad , \quad \gamma_{zx} = \gamma_{xz} = \frac{\partial u_z}{\partial x} + \frac{\partial u_x}{\partial z}$$

The tensorial shear strain components of the infinitesimal strain tensor can then be expressed using the engineering strain definition, γ, as:

$$\underline{\underline{\varepsilon}} = \begin{bmatrix} \varepsilon_{xx} & \varepsilon_{xy} & \varepsilon_{xz} \\ \varepsilon_{yx} & \varepsilon_{yy} & \varepsilon_{yz} \\ \varepsilon_{zx} & \varepsilon_{zy} & \varepsilon_{zz} \end{bmatrix} = \begin{bmatrix} \varepsilon_{xx} & \frac{1}{2}\gamma_{xy} & \frac{1}{2}\gamma_{xz} \\ \frac{1}{2}\gamma_{yx} & \varepsilon_{yy} & \frac{1}{2}\gamma_{yz} \\ \frac{1}{2}\gamma_{zx} & \frac{1}{2}\gamma_{zy} & \varepsilon_{zz} \end{bmatrix}.$$

Metric Tensor

A strain field associated with a displacement is defined, at any point, by the change in length of the tangent vectors representing the speeds of arbitrarily parametrized curves passing through that point. A basic geometric result, due to Fréchet, von Neumann and Jordan, states that, if the lengths of the tangent vectors fulfil the axioms of a norm and the parallelogram law, then the length of a vector is the square root of the value of the quadratic form associated, by the polarization formula, with a positive definite bilinear map called the metric tensor.

Stress–strain Curve

The stress-strain graph provides engineers and designers a graphical measure of the strength and elasticity of a material. It allows them to predict the behavior of materials used in a given application. To draw the graph, the material must first undergo a tensile test.

A tensile test is the simplest and most widely implemented mechanical test. In it, a rod composed of the material to be tested is clamped between two claws. To test its strength and elasticity, the two claws pull the rod at a standard rate in opposite directions to elongate it until it breaks. The rod is subjected to a stress that is equal to the ratio of the opposite forces applied on both ends (P) and the rod's cross-sectional area (A). This stress deforms the rod by elongating it, the rod is subjected to a strain, which is equal to the ratio of change in the length of the rod (δL) and its original length (L).

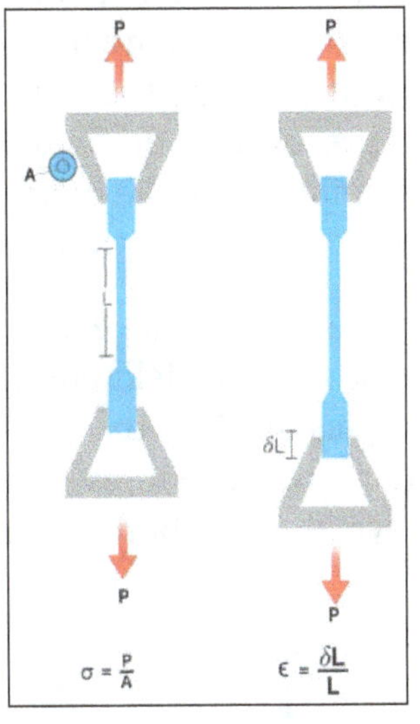

As the stress increases, the strain caused by it varies according to the properties of a material. The relationship can be limned by a graph, and this graph is referred to as the stress-strain curve, where stress is plotted on the Y-axis and strain is plotted on the X-axis. Although there are a plethora of materials at our disposal, we can essentially classify them into two groups. Thus, there are two stress-strain curves, one that describes ductile materials and another that describes brittle materials. Let's first consider the former.

Ductile Curve

When a ductile material, such as copper or aluminum, is put under stress, initially, the resulting strain is proportional to the magnitude of the forces. This is depicted by the straight-line OA. The straight-line implies that stress and strain share a linear or direct relationship throughout OA. Or, simply, the material obeys Hooke's law. The proportionality can be removed by equating them with a constant, in this case, Young's modulus or the modulus of elasticity of the material. Stress 'σ' is then equal to the product of Young's modulus 'E' and strain '∈'. Furthermore, in this region, the material behaves like an elastic it retains or recovers its original shape whenever it is unclamped and freed from stress.

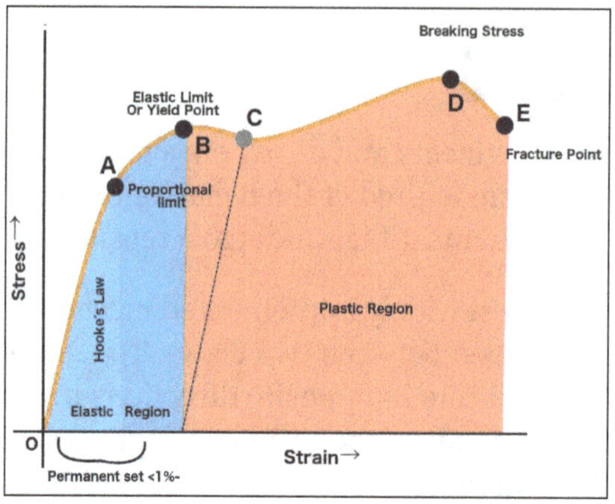

This is also true for region AB, except that the material in this region doesn't obey Hooke's law, a fact one can easily discern from its non-linear shape. Point A is called the proportional limit because beyond this point, stress and strain cease to share their linear relationship. Still, even though the material doesn't obey Hooke's Law, it manages to retain its elasticity. If you were to unclamp the rod in AB, the material would still recover its original shape. Point B is therefore known as the elastic point, it represents the maximum force the material can withstand elastically, and the region OB is called the elasticity region.

Further stress, however, pushes the material over the elastic point's edge to the plasticity region. This is why Point B is also known as the uppper-yield point, as beyond it the material succumbs to stress and begins to deform, like a plastic. Here the stress suddenly dips, but the strain is so severe that the deformation—more precisely, elongation—is permanent. The stress decreases until point C, which is known as the lower-yield point, but as one can observe, the material continues to elongate.

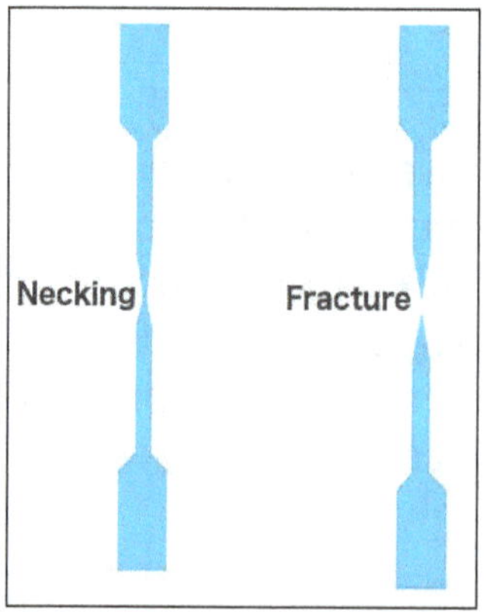

When plasticity is induced in the material, its internal molecular structure suffers constant rearrangements. The material tries to resist this change and tends to harden. This is known as strain hardening. However, as the stress applied to it increases, it continues to elongate along its length, progressively growing longer and thinner, until point D, which represents the material's maximum strength. This is referred to as the material's ultimate strength point.

Stress greater than D is so excruciating that the elongation causes the formation of a neck at the rod's weakest point. Now, only the neck experiences any further deformation. The neck grows thinner and weaker until point E, where it breaks. The material is now said to be fractured. The two broken parts exhibit what is called a 'cup and cone' geometry. Point E is called the fracture or rupture point.

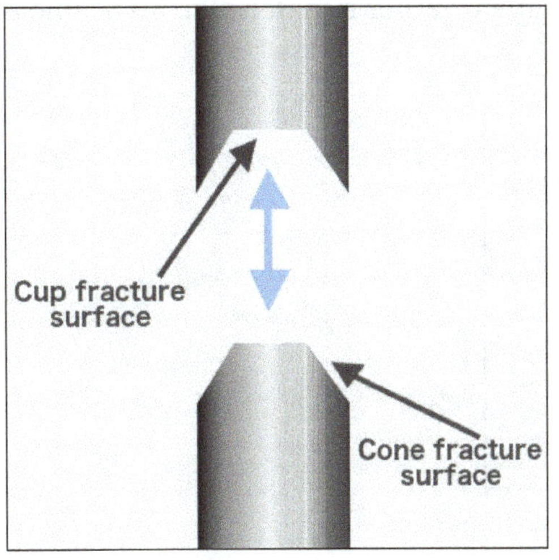

The stress-strain curve of every ductile material may have different values of A, B, C, D, and E, but all the curves generally draw an eerily similar trajectory.

Brittle Curve

Brittle materials such as glass share the same relationship except they don't exhibit a yielding phenomenon. These materials skip the plasticity region and undergo a fracture directly after the elasticity region. The major difference between ductile and brittle materials is that while the former experiences extensive deformation, the latter suffers very little deformation. It survives until the elastic limit, after which it fractures when subjected to excess stress. Glass is the most palpable example; since it cannot be molded (until it is heated, as stress-strain curves are subject to changes in temperature), it can only be fractured.

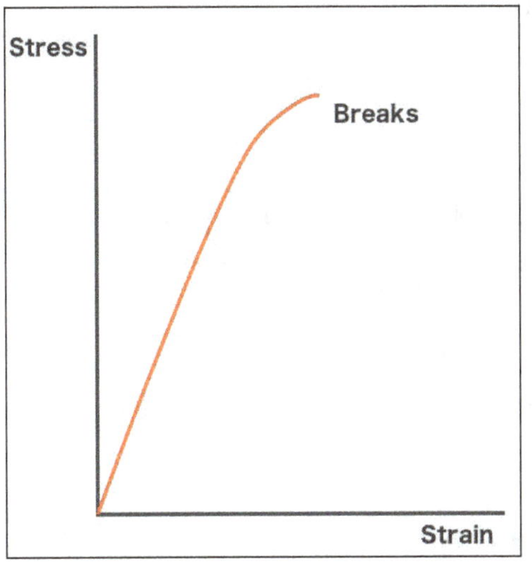

Such a relationship can be limned for every material. If we were to widen our scope beyond ductile and brittle materials, we'd be surprised to find materials such as plastics that don't even have a fracture point. However, its proportional limit is negligible. On the contrary, steel, whose graph displays a steep initial slope, boasts a healthy proportional limit that indicates its strength and durability, but poor plasticity, as its fracture point is not far away. One can then examine the graph of various materials to determine whether a material is malleable enough to be used as a wire or rigid enough to build a bridge.

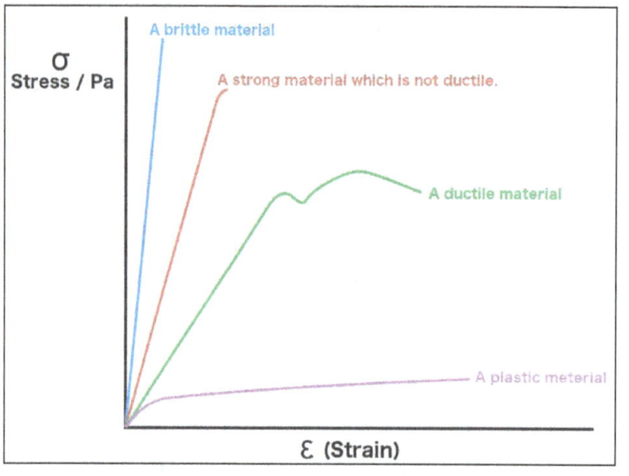

Poisson's Ratio

In mechanics, Poisson's ratio is the negative of the ratio of transverse strain to lateral or axial strain. It is named after Siméon Poisson. It is the ratio of the amount of transversal expansion to the amount of axial compression for small values of these changes.

Poisson's ratio is "the ratio of transverse contraction strain to longitudinal extension strain in the direction of the stretching force." Here,

- Compressive deformation is considered negative.
- Tensile deformation is considered positive.

Symbol	Greek letter 'nu',ν
Formula	Poisson's ratio = – Lateral strain/Longitudinal strain
Range	-1.0 to +0.5
Units	Unitless quantity
Scalar/Vector	Scalar quantity

Poisson's Ratio Formula

Imagine a piece of rubber, in the usual shape of a cuboid. Then imagine pulling it along the sides. What happens now?

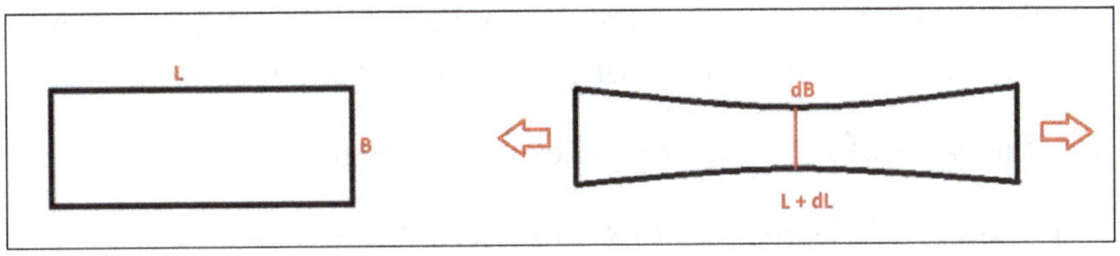

It will compress in the middle. If the original length and breadth of the rubber are taken as L and B respectively, then when pulled longitudinally, it tends to get compressed laterally. In simple words, length has increased by an amount dL and the breadth has increased by an amount dB.

In this case,

$$\varepsilon_t = -\frac{dB}{B}$$

$$\varepsilon_l = -\frac{dL}{L}$$

The formula for Poisson's ratio is,

$$\text{Poisson's ratio} = \frac{\text{Transverse starin}}{\text{Longitudinal strain}} \Rightarrow \nu = -\frac{\varepsilon_t}{\varepsilon_l}$$

where,

εt is the Lateral or Transverse Strain.

εl is the Longitudinal or Axial Strain.

ν is the Poisson's Ratio.

The strain on its own is defined as the change in dimension (length, breadth, area) divided by the original dimension.

Poisson Effect

When a material is stretched in one direction, it tends to compress in the direction perpendicular to that of force application and vice versa. The measure of this phenomenon is given in terms of Poisson's ratio. For example, a rubber band tends to become thinner when stretched.

Poisson's Ratio Values for Different Material

It is the ratio of transverse contraction strain to longitudinal extension strain, in the direction of the stretching force. There can be a stress and strain relation which is generated with the application of force on a body.

- For tensile deformation, Poisson's ratio is positive.
- For compressive deformation, it is negative.

Here, the negative Poisson ratio suggests that the material will exhibit a positive strain in the transverse direction, even though the longitudinal strain is positive as well.

For most materials, the value of Poisson's ratio lies in the range, 0 to 0.5.

A few examples of Poisson ratio is given below for different materials.

Material	Values
Concrete	0.1 – 0.2
Cast iron	0.21 – 0.26
Steel	0.27 – 0.30
Rubber	0.4999
Gold	0.42 – 0.44
Glass	0.18 – 0.3
Cork	0.0
Copper	0.33
Clay	0.30 – 0.45
Stainless steel	0.30 – 0.31
Foam	0.10 – 0.50

Elasticity

Elasticity is a measure of how much an object deforms (strain) when a given stress (force) is applied.

If a bulldozer pushes a car into a wall, the car will not move once it hits the wall, but it will noticeably change shape. A change in shape due to the application of a force is a deformation. Even very small forces are known to cause some deformation. For small deformations, two important characteristics are observed. First, the object returns to its original shape when the force is removed—that is, the deformation is elastic for small deformations. Second, the size of the deformation is proportional to the force—that is, for small deformations, Hooke's law is obeyed. In equation form, Hooke's law is given by $F=k\Delta L$, where ΔL is the change in length.

Elasticity is a measure of how difficult it is to stretch an object. In other words it is a measure of how small kk is. Very elastic materials like rubber have small kk and thus will stretch a lot with only a small force. Stress is a measure of the force put on the object over the area. Strain is the change in length divided by the original length of the object.

Experiments have shown that the change in length (ΔL) depends on only a few variables. As already noted, ΔL is proportional to the force F and depends on the substance from which the object is made. Additionally, the change in length is proportional to the original length Lo and inversely proportional to the cross-sectional area of the wire or rod. For example, a long guitar string will stretch more than a short one, and a thick string will stretch less than a thin one.

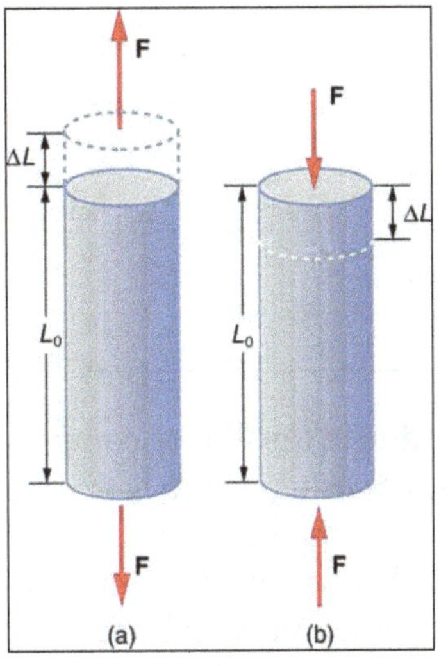

Tension/Compression: Tension: The rod is stretched a length ΔL when a force is applied parallel to its length. (b) Compression: The same rod is compressed by forces with the same magnitude in the opposite direction. For very small deformations and uniform materials, ΔL is approximately the same for the same magnitude of tension or compression. For larger deformations, the cross-sectional area changes as the rod is compressed or stretched.

Fracture

Fracture is caused by a strain placed on an object such that it deforms beyond its elastic limit and breaks.

Materials cannot stretch forever. When a strain is applied to a material it deforms elastically proportional to the force applied. However, after it has deformed a certain amount, the object can no longer take the strain and will break or fracture. The zone in which it bends under strain is called the elastic region. In that region the object will bend and then return to its original shape when the force is abated. Past that point, if more strain is added, the object may permanently deform and eventually fracture.

Fracture strength, also known as breaking strength, is the stress at which a specimen fails via fracture. This is usually determined for a given specimen by a tensile test, which charts the stress-strain curve. The final recorded point is the fracture strength.

Fracture: This is a graph of deformation ΔL versus applied force F. The straight segment is the linear region where Hooke's law is obeyed. The slope of the straight region is 1k. For larger forces, the graph is curved but the deformation is still elastic—L will return to zero if the force is removed. Still greater forces permanently deform the object until it finally fractures. The shape of the curve near fracture depends on several factors, including how the force F is applied. Note that in this graph the slope increases just before fracture, indicating that a small increase in F is producing a large increase in L near the fracture.

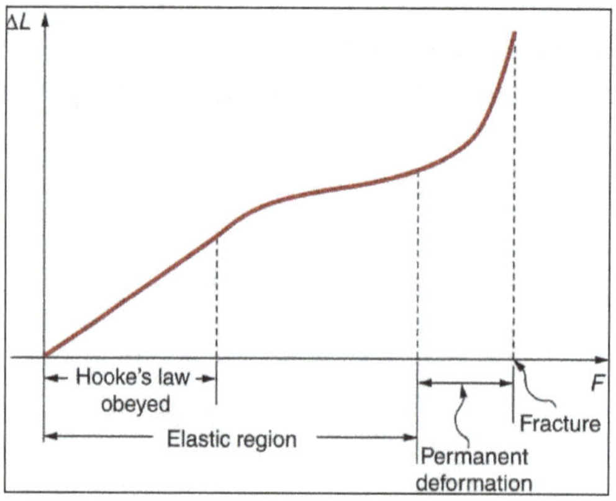

Bones, on the whole, do not fracture due to tension or compression. Rather they generally fracture due to sideways impact or bending, resulting in the bone shearing or snapping. The behavior of bones under tension and compression is important because it determines the load the bones can carry. Bones are classified as weight-bearing structures such as columns in buildings and trees. Weight-bearing structures have special features; columns in building have steel-reinforcing rods while trees and bones are fibrous. The bones in different parts of the body serve different structural functions and are prone to different stresses. Thus, the bone in the top of the femur is arranged in thin sheets separated by marrow while, in other places, the bones can be cylindrical and filled with marrow or just solid. Overweight people have a tendency toward bone damage due to sustained compressions in bone joints and tendons.

Hooke's Law

Hooke's law of elasticity is an approximation that states that the extension of a spring is directly proportional to the load applied to it.

In mechanics (physics), Hooke's law is an approximation of the response of elastic (i.e., springlike) bodies. It states: the extension of a spring is in direct proportion with the load applied to it. For instance, the spring is pulled downwards with either no load, F_p, or twice F_p.

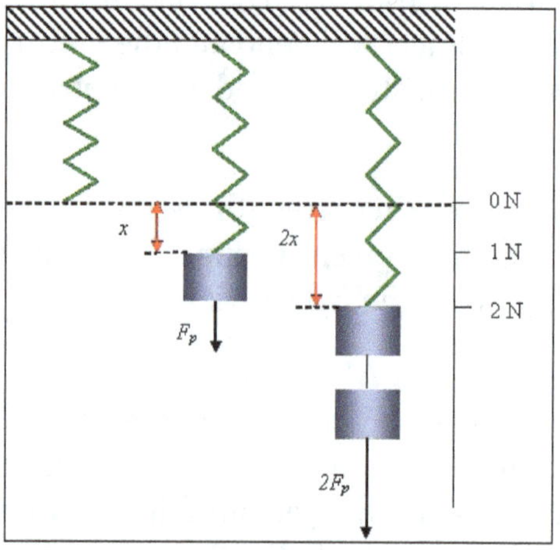

Diagram of Hooke's Law: The extension of the spring is linearly proportional to the force.

Springs and Hooke's Law

Many materials obey this law of elasticity as long as the load does not exceed the material's elastic limit. Materials for which Hooke's law is a useful approximation are known as linear-elastic or "Hookean" materials. Hookean materials are broadly defined and include springs as well as muscular layers of the heart. In simple terms, Hooke's law says that stress is directly proportional to strain. Mathematically, Hooke's law is stated as:

$$F = -kx.$$

where:

- x is the displacement of the spring's end from its equilibrium position (a distance, in SI units: meters);
- F is the restoring force exerted by the spring on that end (in SI units: N or $kg \cdot m/s^2$); and
- k is a constant called the rate or spring constant (in SI units: N/m or kg/s^2). When this holds, the behavior is said to be linear. If shown on a graph, the line should show a direct variation.

It's possible for multiple springs to act on the same point. In such a case, Hooke's law can still be applied. As with any other set of forces, the forces of many springs can be combined into one resultant force.

When Hooke's law holds, the behavior is linear; if shown on a graph, the line depicting force as a function of displacement should show a direct variation. There is a negative sign on the right hand side of the equation because the restoring force always acts in the opposite direction of the displacement (for example, when a spring is stretched to the left, it pulls back to the right).

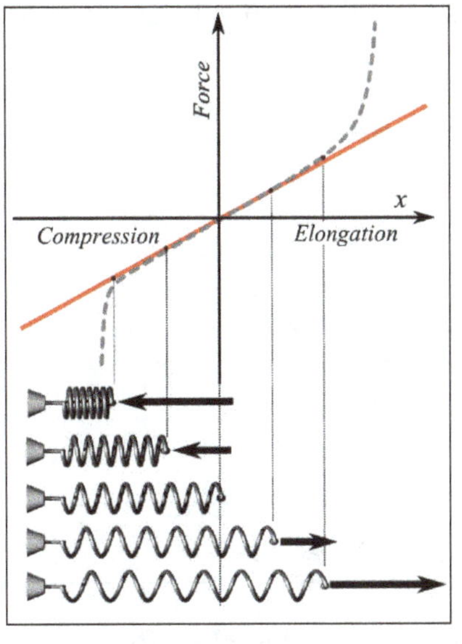

Hooke's Law: The red line in this graph illustrates how force, F, varies with position according to Hooke's law. The slope of this line corresponds to the spring constant k. The dotted line shows what the actual (experimental) plot of force might look like. The pictures of spring states at the bottom of the graph correspond to some points of the plot; the middle one is in the relaxed state (no force applied).

Hooke's law is named after the 17th century British physicist Robert Hooke, and was first stated in 1660 as a Latin anagram, whose solution Hooke published in 1678 as Ut tensio, sic vis, meaning, "As the extension, so the force."

Elastic Potential Energy

If a force results in only deformation, with no thermal, sound, or kinetic energy, the work done is stored as elastic potential energy.

In order to produce a deformation, work must be done. That is, a force must be exerted through a distance, whether you pluck a guitar string or compress a car spring. If the only result is deformation and no work goes into thermal, sound, or kinetic energy, then all the work is initially stored in the deformed object as some form of potential energy. Elastic energy is the potential mechanical energy stored in the configuration of a material or physical system when work is performed to distort its volume or shape. For example, the potential energy PE_{el} stored in a spring is,

$$PE_{el} = \frac{1}{2}kx^2$$

where k is the elastic constant and x is the displacement.

It is possible to calculate the work done in deforming a system in order to find the energy stored. This work is performed by an applied force F_{app}. The applied force is exactly opposite to the restoring force (action-reaction), and so F_{app} = kx. A graph shows the applied force versus deformation x for a system that can be described by Hooke's law. Work done on the system is force multiplied by distance, which equals the area under the curve, or $= \frac{1}{2}kx^2$ Another way to determine the work is to note that the force increases linearly from 0 to kx, so that the average force is $= \frac{1}{2}kx^2$, the distance moved is x.

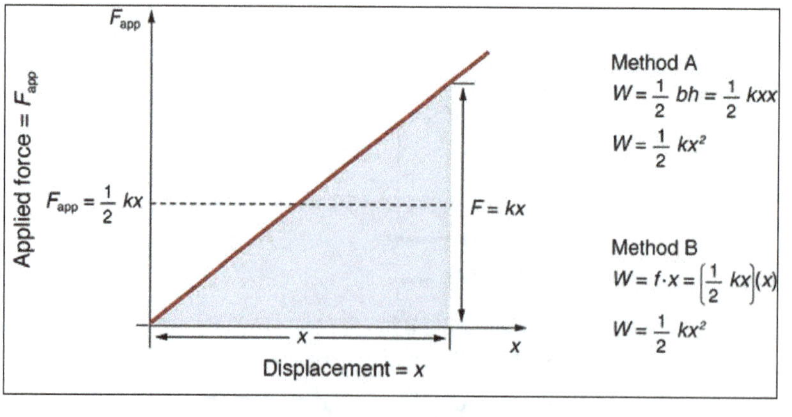

Applied force versus deformation: A graph of applied force versus distance for the deformation of a system that can be described by Hooke's law is displayed. The work done on the system equals the area under the graph or the area of the triangle, which is half its base multiplied by its height, or $W = \frac{1}{2}kx^2$,

$$W = F_{app} d = \left(\frac{1}{2}kx\right)(x) = \frac{1}{2}kx^2.$$

Elastic energy of or within a substance is static energy of configuration. It corresponds to energy stored principally by changing the inter-atomic distances between nuclei. Thermal energy is the randomized distribution of kinetic energy within the material, resulting in statistical fluctuations of the material about the equilibrium configuration. There is some interaction, however. For example, for some solid objects, twisting, bending, and other distortions may generate thermal energy, causing the material's temperature to rise. This energy can also produce macroscopic vibrations sufficiently lacking in randomization to lead to oscillations that are merely the exchange between (elastic) potential energy within the object and the kinetic energy of motion of the object as a whole.

Young's Modulus or Modulus of Elasticity

The modulus of elasticity (= Young's modulus) E is a material property, that describes its stiffness and is therefore one of the most important properties of solid materials.

Mechanical deformation puts energy into a material. The energy is stored elastically or dissipated plastically. The way a material stores this energy is summarized in stress-strain curves. Stress is defined as force per unit area and strain as elongation or contraction per unit length.

When a material deforms elastically, the amount of deformation likewise depends on the size of the material, but the strain for a given stress is always the same and the two are related by Hooke´s Law (stress is directly proportional to strain):

$$\sigma = E.\varepsilon$$

where σ is stress [MPa],

E = modulus of elasticity [MPa]

ε = strain [unitless or %]

From the Hook's law the modulus of elasticity is defined as the ratio of the stress to the strain:

$$E = \frac{\sigma}{\varepsilon}[MPa]$$

Stress is not directly measurable. We can calculate it from different formulas for different types of the loading (tension, flexural stress).

Strain is defined as the change of the length divided by the original (initial) length:

$$\varepsilon = \frac{\Delta I}{I_0} = \frac{I_1 - I_0}{I_0} \ [unitless\ or\ \%]$$

where, Δl is change of the length [m],

l_1 length after elongation [m],

l_o original (initial) length [m].

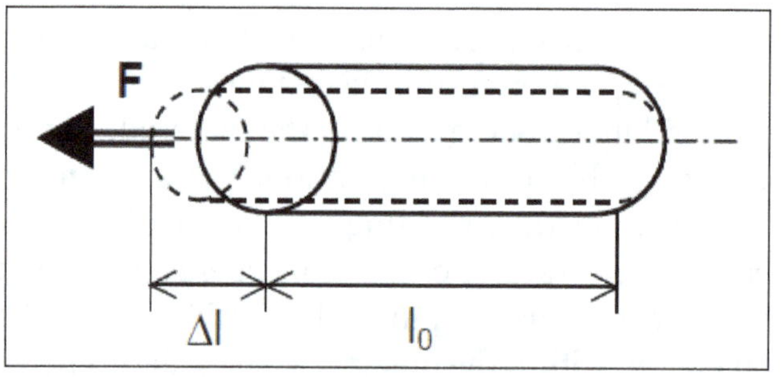

Determination of the strain

Graphical Determination of Modulus of Elasticity

Graphically we can define modulus of elasticity as a slope of the linear portion of the stress-strain diagram.

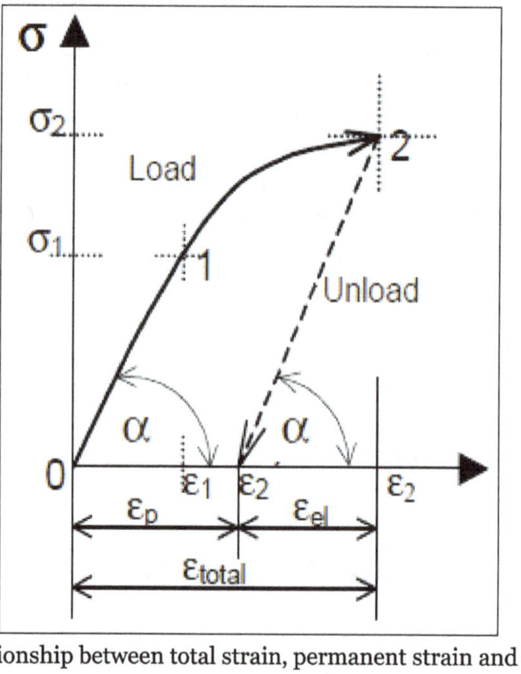

Graphical relationship between total strain, permanent strain and elastic strain

$$E = tg\ \alpha$$

ε_p ......permanent strain

ε_{el} .....elastic strain

Linear portion of stress strain diagram,

$$E = E_1 = \frac{\sigma_1}{\varepsilon_1}$$

Plastic regime of the material,

$$E = E_2 = \frac{\sigma_1}{\varepsilon_1 - \varepsilon_2}$$

The line, describing unloading process of material is parallel with the linear part of stress-strain diagram. It is evident from this, that we are able to determine modulus of elasticity even if the loading is in the unlinear part of the stress-strain diagram. In this case we cannot calculate with whole measured deformation (because part of the deformation is permanent), but we have to unload the material, measure the deformation after unloading and calculate with the difference between deformation in loaded and unloaded stage of material.

Usually we don't know the exact shape of the stress-strain diagram of tested material before measuring and we don't know if the loading is in the linear or unlinear part of the stress-strain diagram. From this reason the process of the test consists of measuring in several loaded and unloaded stages.

Determination of Modulus of Elasticity

The method that have been used to measure modulus of elasticity are following: tension (or compression) test, bending test and natural frequency vibration test. The tension and bending test are based on the principle of Hook's law and they are called static methods. Measuring of the natural frequency of vibration gives dynamic modulus of elasticity.

The static method is based on pulling or bending a sample of the material in an instrument which measures force and measuring the changes of the length. The change of the length is usually very small (tenths – hundredths of mm) and it is impossible to measure it by normal rules. For measuring are used special devices–extensometers. There are several types of extensometers as mechanical or electrical strain gauges.

Modulus of Elasticity in Tension

The test piece is mounted in the tensile testing machine which allows measurable forces to be applied. As extensometer is used mechanical strain gauge. Tensile stress is counted from:

$$\sigma = \frac{F}{A}[MPa]$$

where F is applied force [N]

A cross sectional area [m^2]

Mechanical strain gauge has two contacting point. One point is firm and the second is movable. The distance between points is called gauge (or original) length. Between points is device which measures spacing between them. This device should be a dial where the small straight motion of the point is transmitted into bigger rotary motion of the hand usually there are two hands big hand showing divisions on main scale (centesimal or millesimal) and small hand, showing whole millimetres on auxiliary scale.

The change of the length is count as difference between gauge reading in the state of loading and gauge reading in unloaded state.

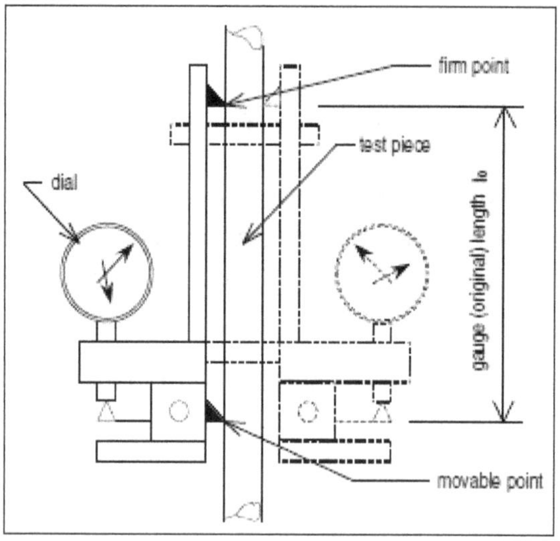

Mechanical strain gauge.

Modulus of Elasticity in Flexural Stress

The test piece is supported at either end and load (in a form of weights) is applied in the middle between the supports. As extensometer is used electrical resistance gauge.

Flexural stress is counted from:

$$\sigma = \frac{M}{W}$$

where, M is bending moment [N. mm²]

W section modulus [mm³].

The principal of electrical resistance gauge is based on the fact that a change in electrical resistance is proportional to the strain, i.e. ΔR/R is function of ε.

Electrical resistance gauge consists of a thin wire grid, fixed on foil (usually paper). The gauge is bonded to the strained surface of measured material by a thin layer of a special glue.

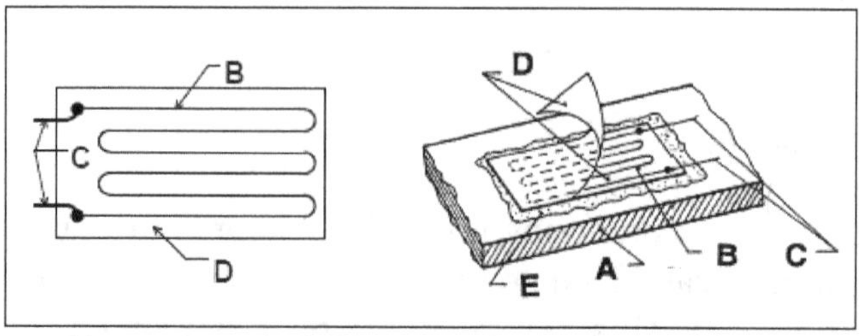

Electrical resistance gauge (a) tested material, (b) wire grid, (c) connected leads, (d) basic foil, (e) glue layer.

The change of resistance is proportional to strain according formula: $\frac{\wedge I}{I} = \frac{1}{K} \times \frac{\wedge R}{R}$

where l is length of the wire,

R electrical resistance,

K gauge factor.

The ideal strain gauge would change resistance only due to the deformations of the surface to which the gauge is attached. However, in real applications, temperature, material properties, the adhesive affect the detected resistance.

Internal influences (as materials properties of wire, type of used glue, size of the wire and the shape of the wire grid) are considered in gauge factor K.

The external influences (the changes of the temperature and moisture) have to be considered in a type of measuring circuit.

In order to measure strain with a bonded resistance strain gauge, it must be connected to an electric circuit that is capable of measuring the small changes in resistance corresponding to strain. Usually the Wheatstone bridge is used, this circuit is also well suited for temperature compensation. Wheatstone bridge is a four arm, four-terminal resistance –measuring network.

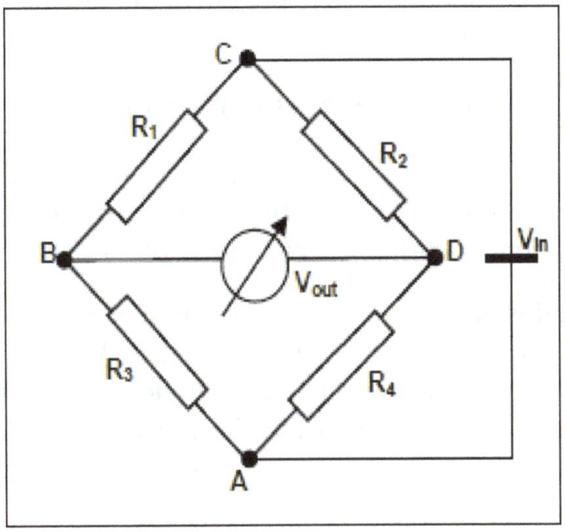

Scheme of Wheatstone bridge.

If the formula: $\frac{R_1}{R_2} = \frac{R_3}{R_4}$.

V_{in} is applied between points A and C, then the output between B and D will show no potential difference = the Wheatstone bridge is balanced (the galvanometer between B and D shows zero).

If R_1 is changed to some value which does not equal R_2, R_3, R_4, the bridge will become unbalanced and a voltage will exist at the output terminals.

Measuring scheme by apparatus TSA is given in Fig Resistance R_1 is a measuring gauge, bonded on tested material in the place of measured deformations. Resistance R_2 is compensating gauge. It is the same type of gauge as R_1, but bonded on non-loaded piece of material. It is used for temperature compensation, but not for strain measurement. The resistance changes due the temperature will have the same value for R_1 and R_2. These changes will cancel, because in the formula for balanced bridge resistances R_1 and R_2 are in ratio. Unbalance of the bridge will than caused only changes of R_1 from deformations.

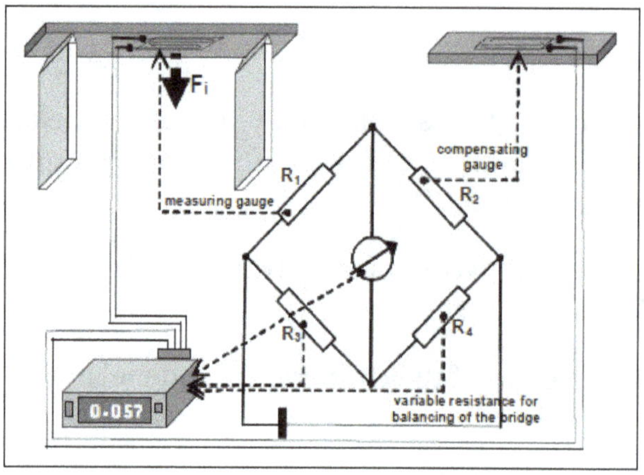

Measuring of modulus of elasticity by the electrical resistant gauge.

Resistances R_3 and R_4 are the components of tensometric apparatus. R_3 is adjustable and serves for nullification of gauge factor. Value of R_3 can be set up in advance. Resistance R_4 is a potentiometer (variable resistance), by it the bridge can be balanced again. The dial of the potentiometer has scale division straight in values of the strain.

Determining of the Basic Strain at Basic Loading

From several reasons it is impossible to measure stress and strain in the low loading range (close to zero loading). Mechanical and electrical machines have lost motions (motion is initiated, but you don't see any results for some time), or there is a risk of the specimen's falling-out from jaws.

If we determine initial modulus of elasticity (i.e. from zero loading to the appropriate loading), we cannot measure deformations straight from zero loading, we begin to measure from some basic loading F_0. We are able to determine basic part of deformations ε_0 (from zero to F_1) from linear part of the stress-strain diagram by or graphically from similarity of triangles.

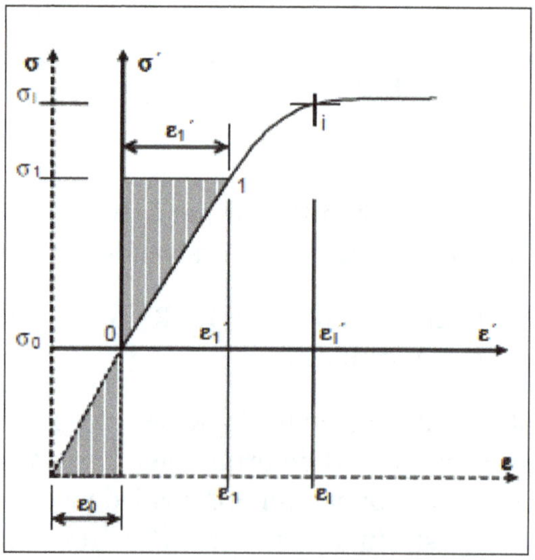

Extrapolation of stress-strain curve in the low loading range.

σ_0 ... basic stress (at F_0)

σ_1 ... stress at loading F_1

ε_1' ... strain at loading F_1

ε_0 ... basic strain (from 0 to F_1).

From similarity of triangles:

$$\varepsilon_0 = \varepsilon'_1 \times \frac{F_0}{F_1 - F_0}$$

Total strain at loading F_i is than:

$$\varepsilon_i = \varepsilon_0 + \varepsilon_i'$$

where ε_i is total strain from zero to appropriate loading F_i

ε_i' strain from F_1 to appropriate loading F_i

ε_0 basic strain from zero to F_1

Bulk Modulus

The bulk modulus is a constant the describes how resistant a substance is to compression. It is defined as the ratio between pressure increase and the resulting decrease in a material's volume. Together with Young's modulus, the shear modulus, and Hooke's law, the bulk modulus describes a material's response to stress or strain.

Usually, bulk modulus is indicated by *K* or *B* in equations and tables. While it applies to uniform compression of any substance, it is most often used to describe the behavior of fluids. It can be used to predict compression, calculate density, and indirectly indicate the types of chemical bonding within a substance. The bulk modulus is considered a descriptor of elastic properties because a compressed material returns to its original volume once the pressure is released.

The units for the bulk modulus are Pascals (Pa) or newtons per square meter (N/m^2) in the metric system, or pounds per square inch (PSI) in the English system.

Table of Fluid Bulk Modulus (K) Values

There are bulk modulus values for solids (e.g., 160 GPa for steel; 443 GPa for diamond; 50 MPa for solid helium) and gases (e.g., 101 kPa for air at constant temperature), but the most common tables list values for liquids. Here are representative values, in both English and metric units:

	English Units (10^5 PSI)	SI Units (10^9 Pa)
Acetone	1.34	0.92
Benzene	1.5	1.05
Carbon Tetrachloride	1.91	1.32

Ethyl Alcohol	1.54	1.06
Gasoline	1.9	1.3
Glycerin	6.31	4.35
ISO 32 Mineral Oil	2.6	1.8
Kerosene	1.9	1.3
Mercury	41.4	28.5
Paraffin Oil	2.41	1.66
Petrol	1.55 - 2.16	1.07 - 1.49
Phosphate Ester	4.4	3
SAE 30 Oil	2.2	1.5
Seawater	3.39	2.34
Sulfuric Acid	4.3	3.0
Water	3.12	2.15
Water - Glycol	5	3.4
Water - Oil Emulsion	3.3	2.3

The K value varies, depending on the state of matter of a sample, and in some cases, on the temperature. In liquids, the amount of dissolved gas greatly impacts the value. A high value of K indicates a material resists compression, while a low value indicates volume appreciably decreases under uniform pressure. The reciprocal of the bulk modulus is compressibility, so a substance with a low bulk modulus has high compressibility.

Upon reviewing the table, you can see the liquid metal mercury is very nearly incompressible. This reflects the large atomic radius of mercury atoms compared with atoms in organic compounds and also the packing of the atoms. Because of hydrogen bonding, water also resists compression.

Bulk Modulus Formulas

The bulk modulus of a material may be measured by powder diffraction, using x-rays, neutrons, or electrons targeting a powdered or microcrystalline sample. It may be calculated using the formula:

Bulk Modulus (*K*) = Volumetric stress /Volumetric strain

This is the same as saying it equals the change in pressure divided by the change in volume divided by initial volume:

Bulk Modulus (*K*) = $(p_1 - p_0) / [(V_1 - V_0) / V_0]$

Here, p_0 and V_0 are the initial pressure and volume, respectively, and p_1 and V1 are the pressure and volume measured upon compression.

Bulk modulus elasticity may also be expressed in terms of pressure and density:

$$K = (p_1 - p_0) / [(\rho_1 - \rho_0) / \rho_0]$$

Here, ρ_0 and ρ_1 are the initial and final density values.

Example:

The bulk modulus may be used to calculate hydrostatic pressure and density of a liquid. For example, consider seawater in the deepest point of the ocean, the Mariana Trench. The base of the trench is 10994 m below sea level.

The hydrostatic pressure in the Mariana Trench may be calculated as:

$$p_1 = \rho * g * h$$

where p_1 is the pressure, ρ is the density of seawater at sea level, g is the acceleration of gravity, and h is the height (or depth) of the water column.

$$p_1 = (1022 \text{ kg/m}^3)(9.81 \text{ m/s}^2)(10994 \text{ m})$$

$$p_1 = 110 \times 10^6 \text{ Pa or } 110 \text{ MPa}$$

Knowing the pressure at sea level is 10^5 Pa, the density of the water at the bottom of the trench may be calculated:

$$\rho_1 = [(p_1 - p)\rho + K*\rho) / K$$

$$\rho_1 = [[(110 \times 10^6 \text{ Pa}) - (1 \times 10^5 \text{ Pa})](1022 \text{ kg/m}^3)] + (2.34 \times 10^9 \text{ Pa})(1022 \text{ kg/m}^3)/(2.34 \times 10^9 \text{ Pa})$$

$$\rho_1 = 1070 \text{ kg/m}^3.$$

What can you see from this? Despite the immense pressure on water at the bottom of the Mariana Trench, it isn't compressed very much.

Rigidity Modulus or Shear Modulus

The shear modulus is defined as the ratio of shear stress to shear strain. It is also known as the modulus of rigidity and may be denoted by *G* or less commonly by *S* or μ. The SI unit of shear modulus is the Pascal (Pa), but values are usually expressed in gigapascals (GPa). In English units, shear modulus is given in terms of pounds per square inch (PSI) or kilo (thousands) pounds per square in (ksi).

- A large shear modulus value indicates a solid is highly rigid. In other words, a large force is required to produce deformation.
- A small shear modulus value indicates a solid is soft or flexible. Little force is needed to deform it.
- One definition of a fluid is a substance with a shear modulus of zero. Any force deforms its surface.

Shear Modulus Equation

The shear modulus is determined by measuring the deformation of a solid from applying a force parallel to one surface of a solid, while an opposing force acts on its opposite surface and holds the solid in place. Think of shear as pushing against one side of a block, with friction as the opposing force. Another example would be attempting to cut wire or hair with dull scissors.

The equation for the shear modulus is:

$$G = \tau_{xy} / \gamma_{xy} = F/A / \Delta x/l = Fl / A\Delta x.$$

Where:

- G is the shear modulus or modulus of rigidity.
- τ_{xy} is the shear stress.
- γ_{xy} is the shear strain.
- A is the area over which the force acts.
- Δx is the transverse displacement.
- l is the initial length.

Shear strain is $\Delta x/l = \tan\theta$ or sometimes $= \theta$, where θ is the angle formed by the deformation produced by the applied force.

Example:

For example, find the shear modulus of a sample under a stress of 4×10^4 N/m² experiencing a strain of 5×10^{-2}.

$$G = \tau / \gamma = (4 \times 10^4 \text{ N/m}^2) / (5 \times 10^{-2}) = 8 \times 10^5 \text{ N/m}^2 \text{ or } 8 \times 10^5 \text{ Pa} = 800 \text{ KPa}.$$

Isotropic and Anisotropic Materials

Some materials are isotropic with respect to shear, meaning the deformation in response to a force is the same regardless of orientation. Other materials are anisotropic and respond differently to stress or strain depending on orientation. Anisotropic materials are much more susceptible to shear along one axis than another. For example, consider the behavior of a block of wood and how it might respond to a force applied parallel to the wood grain compared to its response to a force applied perpendicular to the grain. Consider the way a diamond responds to an applied force. How readily the crystal shears depends on the orientation of the force with respect to the crystal lattice.

Effect of Temperature and Pressure

As you might expect, a material's response to an applied force changes with temperature and pressure. In metals, shear modulus typically decreases with increasing temperature. Rigidity decreases with increasing pressure. Three models used to predict the effects of temperature and pressure on shear modulus are the Mechanical Threshold Stress (MTS) plastic flow stress model, the Nadal and LePoac (NP) shear modulus model, and the Steinberg-Cochran-Guinan (SCG) shear modulus model. For metals, there tends to be a region of temperature and pressures over which change in shear modulus is linear. Outside of this range, modeling behavior is trickier.

Table of Shear Modulus Values

This is a table of sample shear modulus values at room temperature. Soft, flexible materials tend to have low shear modulus values. Alkaline earth and basic metals have intermediate values. Transition metals and alloys have high values. Diamond, a hard and stiff substance, has an extremely high shear modulus.

Material	Shear Modulus (GPa)
Rubber	0.0006
Polyethylene	0.117
Plywood	0.62
Nylon	4.1
Lead (Pb)	13.1
Magnesium (Mg)	16.5
Cadmium (Cd)	19
Kevlar	19
Concrete	21
Aluminum (Al)	25.5
Glass	26.2
Brass	40
Titanium (Ti)	41.1
Copper (Cu)	44.7
Iron (Fe)	52.5
Steel	79.3
Diamond (C)	478.0

Note that the values for Young's modulus follow a similar trend. Young's modulus is a measure of a solid's stiffness or linear resistance to deformation. Shear modulus, Young's modulus, and bulk modulus are modulii of elasticity, all based on Hooke's law and connected to each other via equations.

Plasticity

Plasticity is the ability of certain solids to flow or to change shape permanently when subjected to stresses of intermediate magnitude between those producing temporary deformation, or elastic behaviour, and those causing failure of the material, or rupture. Plasticity enables a solid under the action of external forces to undergo permanent deformation without rupture. Elasticity, in comparison, enables a solid to return to its original shape after the load is removed. Plastic deformation occurs in many metal-forming processes (rolling, pressing, forging) and in geologic processes (rock folding and rock flow within the earth under extremely high pressures and at elevated temperatures).

Plastic deformation is a property of ductile and malleable solids. Brittle materials, such as cast iron, cannot be plastically deformed, though at elevated temperatures some, such as glass, which is not a crystallized solid, do undergo plastic flow.

Plasticity, as the name of a science, refers either to mathematical descriptions of what happens in plastic deformation in terms of stresses, strains, and loads or to physical explanations of plastic flow in terms of atoms, crystals, grains, and motions of structural defects (dislocations) within crystals.

The classical theory of plasticity grew out of the study of metals in the late nineteenth century. It is concerned with materials which initially deform elastically, but which deform plastically upon reaching a yield stress. In metals and other crystalline materials the occurrence of plastic deformations at the micro-scale level is due to the motion of dislocations and the migration of grain boundaries on the micro-level. In sands and other granular materials plastic flow is due both to the irreversible rearrangement of individual particles and to the irreversible crushing of individual particles. Similarly, compression of bone to high stress levels will lead to particle crushing. The deformation of microvoids and the development of micro-cracks is also an important cause of plastic deformations in materials such as rocks.

There are two broad groups of metal plasticity problem which are of interest to the engineer and analyst. The first involves relatively small plastic strains, often of the same order as the elastic strains which occur. Analysis of problems involving small plastic strains allows one to design structures optimally, so that they will not fail when in service, but at the same time are not stronger than they really need to be. In this sense, plasticity is seen as a material failure.

The second type of problem involves very large strains and deformations, so large that the elastic strains can be disregarded. These problems occur in the analysis of metals manufacturing and forming processes, which can involve extrusion, drawing, forging, rolling and so on. In these latter-type problems, a simplified model known as perfect plasticity is usually employed, and use is made of special limit theorems which hold for such models.

Plastic deformations are normally rate independent, that is, the stresses induced are independent of the rate of deformation (or rate of loading). This is in marked contrast to classical Newtonian fluids for example, where the stress levels are governed by the rate of deformation through the viscosity of the fluid.

Materials commonly known as "plastics" are not plastic in the sense described here. They, like other polymeric materials, exhibit viscoelastic behaviour where, as the name suggests, the material response has both elastic and viscous components. Due to their viscosity, their response is, unlike the plastic materials, rate-dependent. Further, although the viscoelastic materials can suffer irrecoverable deformation, they do not have any critical yield or threshold stress, which is the characteristic property of plastic behaviour. When a material undergoes plastic deformations, i.e. irrecoverable and at a critical yield stress and these effects are rate dependent, the material is referred to as being viscoplastic.

Observations from Standard Tests

Tension Test

Consider the following key experiment, the tensile test, in which a small, usually cylindrical, specimen is gripped and stretched, usually at some given rate of stretching. The force required to hold the specimen at a given stretch is recorded If the material is a metal, the deformation remains elastic up to a certain force level, the yield point of the material. Beyond this point, permanent plastic deformations are induced. On unloading only the elastic deformation is recovered and the specimen will have undergone a permanent elongation (and consequent lateral contraction).

In the elastic range the force-displacement behaviour for most engineering materials (metals, rocks, plastics, but not soils) is linear. After passing the elastic limit, the material "gives" and is said to undergo plastic flow. Further increases in load are usually required to maintain the plastic flow and an increase in displacement; this phenomenon is known as work-hardening or strain-hardening. In some cases, after an initial plastic flow and hardening, the force-displacement curve decreases, as in some soils; the material is said to be softening. If the specimen is unloaded from a plastic state (B) it will return along the path BC shown, parallel to the original elastic line. This is elastic recovery. The strain which remains upon unloading is the permanent plastic deformation. If the material is now loaded again, the force-displacement curve will retrace the unloading path CB until it again reaches the plastic state. Further increases in stress will cause the curve to follow BD.

Two important observations concerning the above tension test (on most metals) are the following:

(1) After the onset of plastic deformation, the material will be seen to undergo negligible volume change, that is, it is incompressible.

(2) The force-displacement curve is more or less the same regardless of the rate at which the specimen is stretched (at least at moderate temperatures).

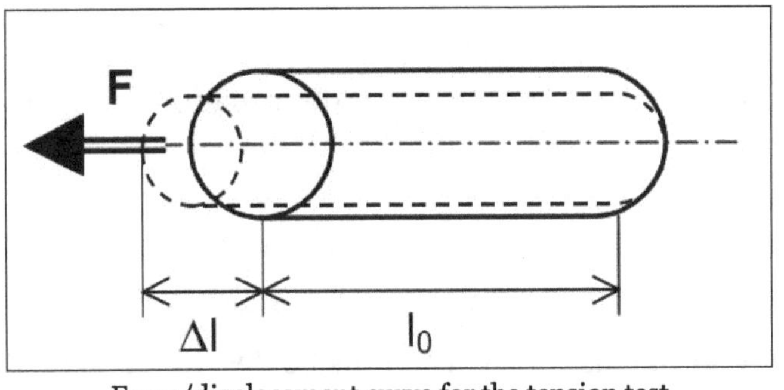

Force/displacement curve for the tension test

Nominal and True Stress and Strain

There are two different ways of describing the force F which acts in a tension test. First, normalising with respect to the original cross sectional area of the tension test specimen A_0, one has the nominal stress or engineering stress,

$$\sigma_n = \frac{F}{A_0}$$

Alternatively, one can normalise with respect to the current cross-sectional area A, leading to the true stress,

$$\sigma = \frac{F}{A}$$

in which F and A are both changing with time. For very small elongations, within the elastic range say, the cross-sectional area of the material undergoes negligible change and both definitions of stress are more or less equivalent.

Similarly, one can describe the deformation in two alternative ways. Denoting the original specimen length by l_0 and the current length by l, one has the engineering strain,

$$\varepsilon = \frac{l - l_0}{l_0}$$

Alternatively, the true strain is based on the fact that the "original length" is continually changing; a small change in length dl leads to a strain increment dε = dl/l and the total strain is defined as the accumulation of these increments:

$$\varepsilon_t = \int_{l_0}^{l} \frac{dl}{l} = \ln\left(\frac{l}{l_0}\right)$$

The true strain is also called the logarithmic strain or Hencky strain. Again, at small deformations, the difference between these two strain measures is negligible. The true strain and engineering strain are related through,

$$\varepsilon_t = In(1 + \varepsilon)$$

Using the assumption of constant volume for plastic deformation and ignoring the very small elastic volume changes,

$$\sigma = \sigma_n \frac{l}{l_0}.$$

The stress-strain diagram for a tension test can now be described using the true stress/strain or nominal stress/strain definitions, as in the figure below. A here denotes the point at which the maximum force the specimen can withstand has been reached. The nominal stress at A is called the Ultimate Tensile Strength (UTS) of the material. After this point, the specimen "necks", with a very rapid reduction in cross-sectional area somewhere about the centre of the specimen until the specimen ruptures, as indicated by the asterisk.

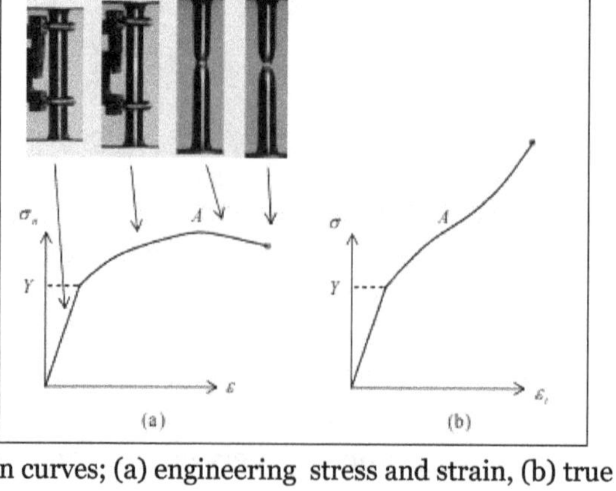

Typical stress/strain curves; (a) engineering stress and strain, (b) true stress and strain.

Note that, during loading into the plastic region, the yield stress increases. For example, if one unloads and re-loads, the material stays elastic up until a stress higher than the original yield stress Y. In this respect, the stress-strain curve can be regarded as a yield stress versus strain curve.

Compression Test

A compression test will lead to similar results as the tensile stress. The yield stress in compression will be approximately the same as (the negative of) the yield stress in tension. If one plots the true stress versus true strain curve for both tension and compression (absolute values for the compression), the two curves will more or less coincide. This would indicate that the behaviour of the material under compression is broadly similar to that under tension. If one were to use the nominal stress and strain, then the two curves would not coincide; this is one of a number of good reasons for using the true definitions.

The Bauschinger Effect

If one takes a virgin sample and loads it in tension into the plastic range, and then unloads it and continues on into compression, one finds that the yield stress in compression is not the same as the yield strength in tension, as it would have been if the specimen had not first been loaded in tension. In fact the yield point in this case will be significantly less than the corresponding yield stress in tension. This reduction in yield stress is known as the Bauschinger effect. The solid line depicts the response of a real material. The dotted lines are two extreme cases which are used in plasticity models; the first is the isotropic hardening model, in which the yield stress in tension and compression are maintained equal, the second being kinematic hardening, in which the total elastic range is maintained constant throughout the deformation.

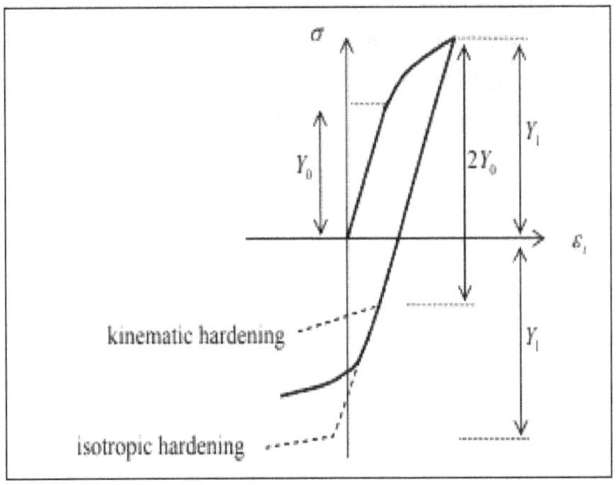

The Bauschinger effect

The presence of the Bauschinger effect complicates any plasticity theory. However, it is not an issue provided there are no reversals of stress in the problem under study.

Hydrostatic Pressure

Careful experiments show that, for metals, the yield behaviour is independent of hydrostatic pressure. That is, a stress state $\sigma_{xx} = \sigma_{yy} = \sigma_{zz} = -p$ has negligible effect on the yield stress of a material, right up to very high pressures. Note however that this is not true for soils or rocks.

Assumptions of Plasticity Theory

Regarding the above test results then, in formulating a basic plasticity theory with which to begin, the following assumptions are usually made:

1. The response is independent of rate effects.
2. The material is incompressible in the plastic range.
3. There is no bauschinger effect.
4. The yield stress is independent of hydrostatic pressure.
5. The material is isotropic.

The first two of these will usually be very good approximations, the other three may or may not be, depending on the material and circumstances. For example, most metals can be regarded as isotropic. After large plastic deformation however, for example in rolling, the material will have become anisotropic: there will be distinct material directions and asymmetries.

Together with these, assumptions can be made on the type of hardening and on whether elastic deformations are significant. For example, consider the hierarchy of models illustrated in figure below, commonly used in theoretical analyses. In (a) both the elastic and plastic curves are assumed linear. In (b) work-hardening is neglected and the yield stress is constant after initial yield. Such perfectly-plastic models are particularly appropriate for studying processes where the metal is worked at a high temperature, such as hot rolling, where work hardening is small. In many areas of applications the strains involved are large, e.g. in metal working processes such as extrusion, rolling or drawing, where up to 50% reduction ratios are common. In such cases the elastic strains can be neglected altogether as in the two models (c) and (d). The rigid/perfectly-plastic model (d) is the crudest of all and hence in many ways the most useful. It is widely used in analysing metal forming processes, in the design of steel and concrete structures and in the analysis of soil and rock stability.

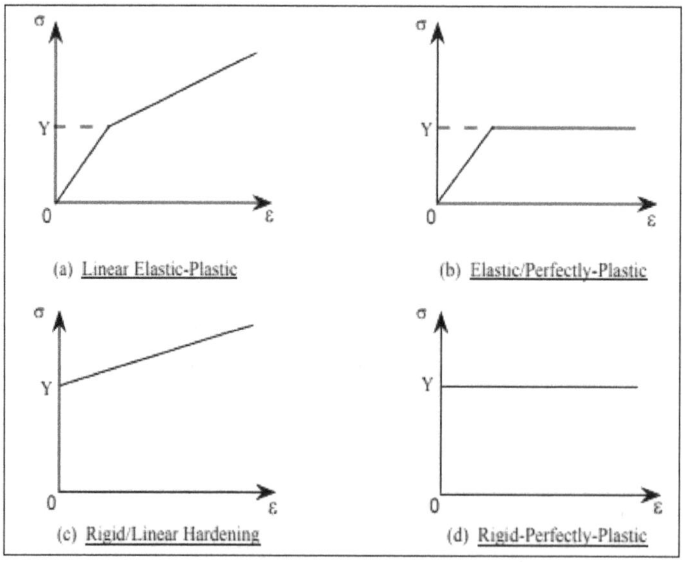

Simple models of elastic and plastic deformation.

The Tangent and Plastic Modulus

Stress and strain are related through σ = Eε in the elastic region, E being the Young's modulus The tangent modulus K is the slope of the stress-strain curve in the plastic region and will in general change during a deformation. At any instant of strain, the increment in stress dσ is related to the increment in strain dε through,

$$d\sigma = Kd\varepsilon.$$

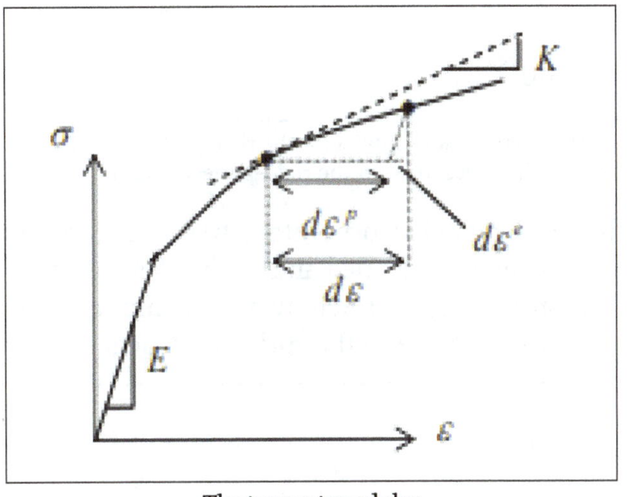

The tangent modulus

After yield, the strain increment consists of both elastic, e^ε , and plastic, p d^ε , strains:

$$d\varepsilon = d\varepsilon^e + d\varepsilon^p$$

The stress and plastic strain increments are related by the plastic modulus H:

$$d\sigma = H\, d\varepsilon^p$$

and it follows that

$$\frac{1}{K} = \frac{1}{E} + \frac{1}{H}.$$

Friction Block Models

Some additional insight into the way plastic materials respond can be obtained from friction block models. The rigid perfectly plastic model can be simulated by a Coulomb friction block. No strain occurs until σ reaches the yield stress Y. Then there is movement – although the amount of movement or plastic strain cannot be determined without more information being available. The stress cannot exceed the yield stress in this model:

$$|\sigma| \le Y$$

If unloaded, the block stops moving and the stress returns to zero, leaving a permanent strain.

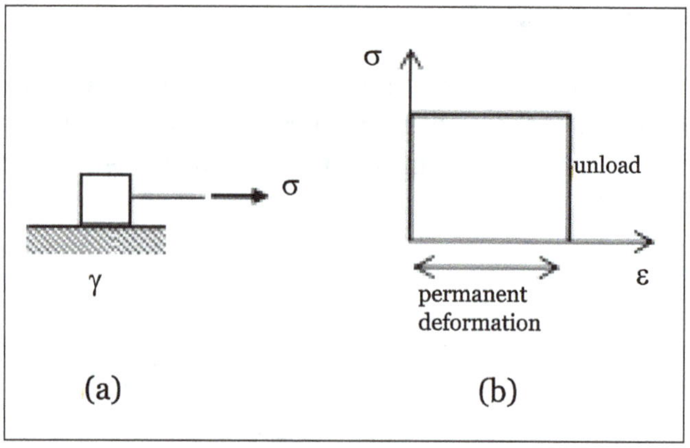

(a) Friction block model for the rigid perfectly plastic material, (b) response of the rigid-perfectly plastic model.

The linear elastic perfectly plastic model incorporates a free spring with modulus E in series with a friction block. The spring stretches when loaded and the block also begins to move when the stress reaches Y, at which time the spring stops stretching, the maximum possible stress again being Y. Upon unloading, the block stops moving and the spring contracts.

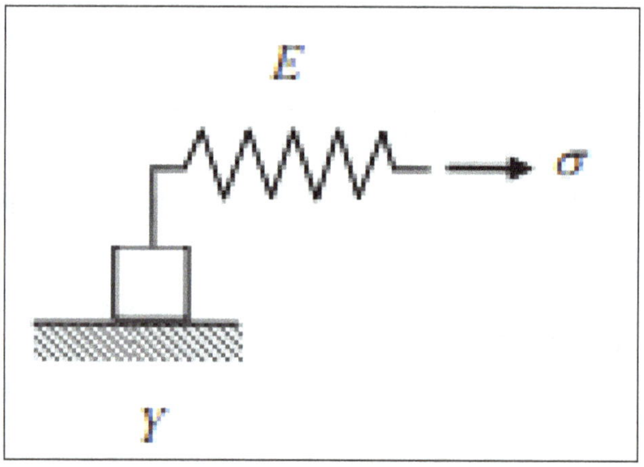

Friction block model for the elastic perfectly plastic material.

The linear elastic plastic model with linear strain hardening incorporates a second, hardening, spring with stiffness H, in parallel with the friction block. Once the yield stress is reached, an ever increasing stress needs to be applied in order to keep the block moving – and elastic strain continues to occur due to further elongation of the free spring. The stress is then split into the yield stress, which is carried by the moving block, and an overstress σ – Y carried by the hardening spring.

Upon unloading, the block "locks" – the stress in the hardening spring remains constant whilst the free spring contracts. At zero stress, there is a negative stress taken up by the friction block, equal and opposite to the stress in the hardening spring.

The slope of the elastic loading line is E. For the plastic hardening line,

$$\varepsilon = \varepsilon^e + \varepsilon^p = \frac{\sigma}{E} + \frac{\sigma - Y}{H} \rightarrow K = \frac{d\sigma}{d\varepsilon} = \frac{EH}{E+H}.$$

It can be seen that H is the plastic modulus.

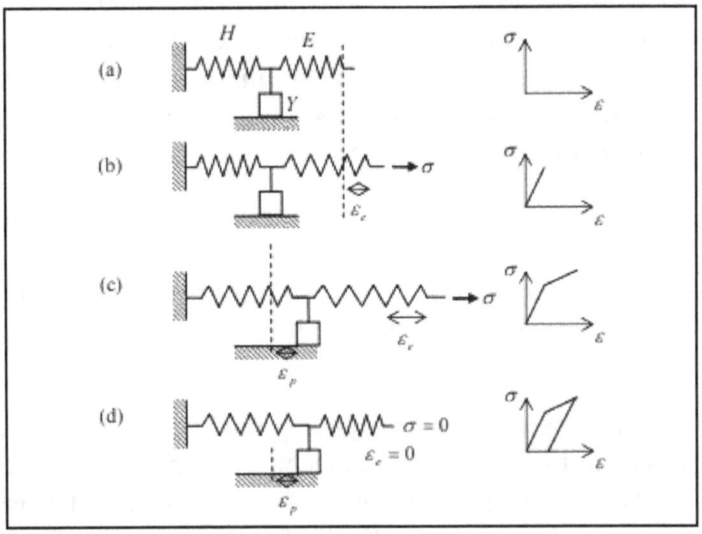

Friction block model for a linear elastic-plastic material with linear strain hardening; (a) stress-free, (b) elastic strain, (c) elastic and plastic strain, (d) unloading.

Yield Strength

The yield point is defined in engineering and materials science as the stress at which a material begins to plastically deform. Prior to the yield point the material will deform elastically and will return to its original shape when the applied stress is removed. Once the yield point is passed some fraction of the deformation will be permanent and non-reversible. Knowledge of the yield point is vital when designing a component since it generally represents an upper limit to the load that can be applied. It is also important for the control of many materials production techniques such as forging, rolling, or pressing.

In structural engineering, yield is the permanent plastic deformation of a structural member under stress. This is a soft failure mode which does not normally cause catastrophic failure unless it accelerates buckling.

It is often difficult to precisely define yield due to the wide variety of stress-strain behaviours exhibited by real materials. In addition there are several possible ways to define the yield point in a given material.

Yeild occurs when dislocations first begin to move. Given that dislocations begin to move at very low stresses, and the difficulty in detecting such movement, this definition is rarely used.

- Elastic Limit - The lowest stress at which permanent deformation can be measured. This requires a complex iteractive load-unload procedure and is critically dependent on the accuracy of the equipment and the skill of the operator.
- Proportional Limit - The point at which the stress-strain curve becomes nonlinear. In most metallic materials the elastic limit and proportional limit are essentially the same.

- Offset Yield Point (proof stress) - Due to the lack of a clear border between the elastic and plastic regions in many materials, the yield point is often defined as the stress at some arbitrary plastic strain (typically 0.2%). This is determined by the intersection of a line offset from the linear region by the required strain. In some materials there is essentially no linear region and so a certain value of plastic strain is defined instead. Although somewhat arbitrary this method does allow for a consistent comparison of materials and is the most common.

Yield point

If the stress is too large, the strain deviates from being proportional to the stress. The point at which this happens is the yield point because there the material yields, deforming permanently (plastically).

Yield stress. Hooke's law is not valid beyond the yield point. The stress at the yield point is called yield stress, and is an important measure of the mechanical properties of materials. In practice, the yield stress is chosen as that causing a permanent strain of 0.002, which called as proof stresss. The yield stress measures the resistance to plastic deformation.

The yield strength is the stress required to produce a small-specified amount of plastic deformation. The usual definition of this property is the offset yield strength determined by the stress corresponding to the intersection of the stressstrain curve and a line parallel to the elastic part of the curve offset by a specified strain. In the United States the offset is usually specified as a strain of 0.2 or 0.1 percent (e = 0.002 or 0.001).

$$R_{p\,0,2} = \frac{P(strainooffet - 0,0002)}{A_0}$$

A good way of looking at offset yield strength is that after a specimen has been loaded to its 0.2 percent offset yield strength and then unloaded it will be 0.2 percent longer than before the test. The offset yield strength is often referred to in Great Britain as the proof stress, where offset values are either 0.1 or 0.5 percent. The yield strength obtained by an offset method is commonly used for design and specification purposes because it avoids the practical difficulties of measuring the elastic limit or proportional limit.

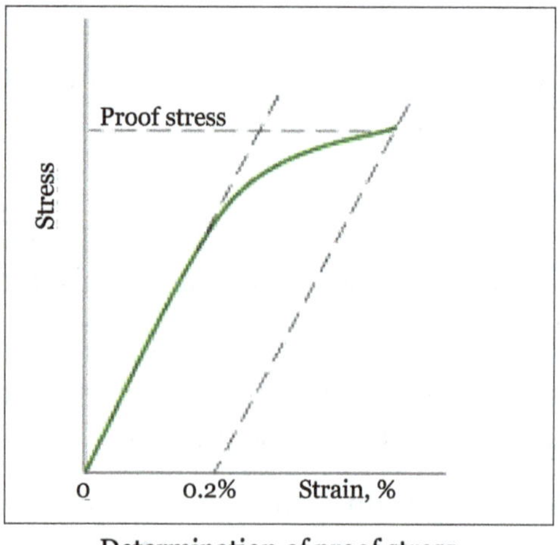

Determination of proof stress.

Some materials have essentially no linear portion to their stress-strain curve, for example, soft copper or gray cast iron. For these materials the offset method cannot be used and the usual practice is to define the yield strength as the stress to produce some total strain, for example, e = 0.005.

Determination of Yield Strength in Ductile Materials

In many materials, the yield stress is not very well defined and for this reason a standard has been developed to determine its value. The standard procedure is to project a line parallel to the initial elastic region starting at 0.002 Strain. The 0.002 strain point is often referred to as 0.2% the offset strain point. The intersection of this new line with the stress-strain curve then defines the yield strength as shown in figure below.

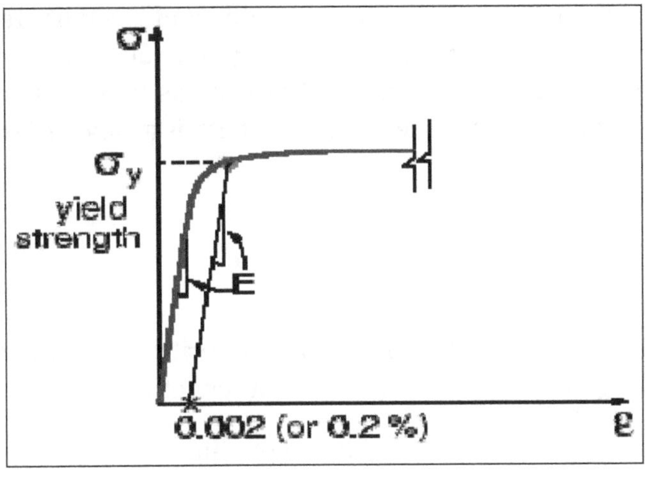

Tensile Strength

When stress continues in the plastic regime, the stress-strain passes through a maximum, called the tensile strength (sTS), and then falls as the material starts to develop a neck and it finally breaks at the fracture point.

Note that it is called strength, not stress, but the units are the same, MPa.

For structural applications, the yield stress is usually a more important property than the tensile strength, since once then it is passed, the structure has deformed beyond acceptable limits.

The tensile strength, or ultimate tensile strength (UTS), is the maximum load divided by the original cross-sectional area of the specimen.

$$R_m = \frac{P_{\max}}{A_0}$$

The tensile strength is the value most often quoted from the results of a tension test; yet in reality it is a value of little fundamental significance with regard to the strength of a metal. For ductile metals the tensile strength should be regarded as a measure of the maximum load, which a metal can withstand under the very restrictive conditions of uniaxial loading. It will be shown that this value bears little relation to the useful strength of the metal under the more complex conditions of stress, which are usually encountered.

For many years it was customary to base the strength of members on the tensile strength, suitably reduced by a factor of safety. The current trend is to the more rational approach of basing the static design of ductile metals on the yield strength.

However, because of the long practice of using the tensile strength to determine the strength of materials, it has become a very familiar property, and as such it is a very useful identification of a material in the same sense that the chemical composition serves to identify a metal or alloy.

Further, because the tensile strength is easy to determine and is a quite reproducible property, it is useful for the purposes of specifications and for quality control of a product. Extensive empirical correlations between tensile strength and properties such as hardness and fatigue strength are often quite useful. For brittle materials, the tensile strength is a valid criterion for design.

Ductility

The ability to deform before braking. It is the opposite of brittleness. Ductility can be given either as percent maximum elongation e_{max} or maximum area reduction.

At our present degree of understanding, ductility is a qualitative, subjective property of a material. In general, measurements of ductility are of interest in three ways:

1. To indicate the extent to which a metal can be deformed without fracture in metal working operations such as rolling and extrusion.
2. To indicate to the designer, in a general way, the ability of the metal to flow plastically before fracture. A high ductility indicates that the material is "forgiving" and likely to deform locally without fracture should the designer err in the stress calculation or the prediction of severe loads.
3. To serve as an indicator of changes in impurity level or processing conditions. Ductility measurements may be specified to assess material quality even though no direct relationship exists between the ductility measurement and performance in service.

The conventional measures of ductility that are obtained from the tension test are the engineering strain at fracture ef (usually called the elongation) and the reduction of area at fracture q. Both of these properties are obtained after fracture by putting the specimen back together and taking measurements of L_f and A_f,

$$e_f = \frac{L_f - L_0}{L_0}$$

$$q = \frac{A_0 - A_f}{A_0}$$

Because an appreciable fraction of the plastic deformation will be concentrated in the necked region of the tension specimen, the value of e_f will depend on the gage length L_0 over which the measurement was taken. The smaller the gage length the greater will be the contribution to the overall elongation from the necked region and the higher will be the value of e_f. Therefore, when reporting values of percentage elongation, the gage length L_0 always should be given.

The reduction of area does not suffer from this difficulty. Reduction of area values can be converted into an equivalent zero-gage-length elongation e_0. From the constancy of volume relationship for plastic deformation $A*L = A_0*L_0$, we obtain:

$$\frac{L}{L_0}=\frac{A_0}{A}=\frac{1}{1-q},\ e_0=\frac{L-L_0}{L_0}=\frac{A_0}{A}-1=\frac{1}{1-q}-1=\frac{q}{1-q}.$$

This represents the elongation based on a very short gage length near the fracture.

Another way to avoid the complication from necking is to base the percentage elongation on the uniform strain out to the point at which necking begins. The uniform elongation e_u correlates well with stretch-forming operations. Since the engineering stress-strain curve often is quite flat in the vicinity of necking, it may be difficult to establish the strain at maximum load without ambiguity. In this case the method suggested by Nelson and Winlock is useful.

Resilience

The resilience of the material is the triangular area underneath the elastic region of the curve. Resilience generally means the ability to recover from (or to resist being affected by) some shock, insult, or disturbance. However, it is used quite differently in different fields.

In physics and engineering, resilience is defined as the capacity of a material to absorb energy when it is deformed elastically and then, upon unloading to have this energy recovered. In other words, it is the maximum energy per volume that can be elastically stored. It is represented by the area under the curve in the elastic region in the Stress-Strain diagram.

Modulus of Resilience, U_r, can be calculated using the following formula:

$$U_r=\frac{\sigma^2}{2E}=0.5\sigma\in=0.5\sigma\left(\frac{\sigma}{E}\right),$$

where σ is yield stress, E is Young's modulus, and ε is strain.

The ability of a material to absorb energy when deformed elastically and to return it when unloaded is called resilience. This is usually measured by the modulus of resilience, which is the strain energy per unit volume required to stress the material from, zero stress to the yield stress.

The ability of a material to absorb energy when deformed elastically and to return it when unloaded is called resilience. This is usually measured by the modulus of resilience, which is the strain energy per unit volume required to stress the material from, zero stress to the yield stress s. The strain energy per unit volume for uniaxial tension is,

$$U_0 = \frac{1}{2}\sigma_x e_x.$$

Table below gives some values of modulus of resilience for different materials.

Table: Modulus of resilience for various materials.

Material	E, psi	s_0, psi	Modulus of resilience,Ur
Medium-carbon steel	30×10^6	45000	33,7
High-carbon spring steel	30×10^6	140000	320
Duraluminium	$10{,}5 \times 10^6$	18000	17,0
Cooper	16×10^6	4000	5,3
Rubber	150	300	300
Acrylic polymer	$0{,}5 \times 10^6$	2000	4,0

Toughness

The area underneath the stress-strain curve is the toughness of the material i.e. the energy the material can absorb prior to rupture.. It also can be defined as the resistance of a material to crack propogation.

In materials science and metallurgy, toughness is the resistance to fracture of a material when stressed. It is defined as the amount of energy that a material can absorb before rupturing, and can be found by finding the area (i.e., by taking the integral) underneath the stress-strain curve.

The ability of a metal to deform plastically and to absorb energy in the process before fracture is termed toughness. The emphasis of this definition should be placed on the ability to absorb energy before fracture. Recall that ductility is a measure of how much something deforms plastically before fracture, but just because a material is ductile does not make it tough. The key to toughness is a good combination of strength and ductility. A material with high strength and high ductility will have more toughness than a material with low strength and high ductility. Therefore, one way to measure toughness is by calculating the area under the stress strain curve from a tensile test. This value is simply called “material toughness” and it has units of energy per volume. Material toughness equates to a slow absorption of energy by the material.

The toughness of a material is its ability to absorb energy in the plastic range. The ability to withstand occasional, stresses above the yield stress without fracturing is particularly desirable in parts such as freight-car couplings, gears, chains, and crane hooks. Toughness is a commonly used concept, which is difficult to pin down and define. One way of looking at toughness is to consider that it is the total area under the stress-strain curve. This area is an indication of the amount of work per unit volume, which can be done, on the material without causing it to rupture. The following Figure shows the stress-strain curves for high- and low-toughness materials. The high-carbon spring steel has a higher yield strength and tensile strength than the medium-carbon structural steel. However, the structural steel is more ductile and has a greater total elongation. The total area under the stresstrain curve is greater for the structural steel, and therefore it is a tougher material. This illustrates that toughness is a parameter that comprises both strength and ductility. The crosshatched regions in Figure indicate the modulus of resilience for each steel. Because of its higher yield strength, the spring steel has the greater resilience.

Several mathematical approximations for the area under the stress-strain curve have been suggested. For ductile metals that have a stress-strain curve like that of the structural steel, the area under the curve can be approximated by either of the following equations:

$$U_T \approx\sim S_u e_f \qquad U_r \approx \frac{S_0 + S_u}{2}.e_f$$

For brittle materials the stress-strain curve is sometimes assumed to be a parabola, and the area under the curve is given by,

$$U_T \approx \frac{2}{3} S_u e_f$$

All these relations are only approximations to the area under the stress-strain curves. Further, the curves do not represent the true behavior in the plastic range, since they are all based on the original area of the specimen.

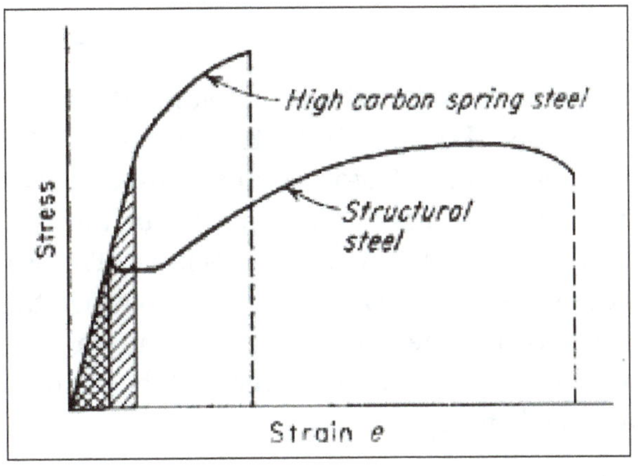

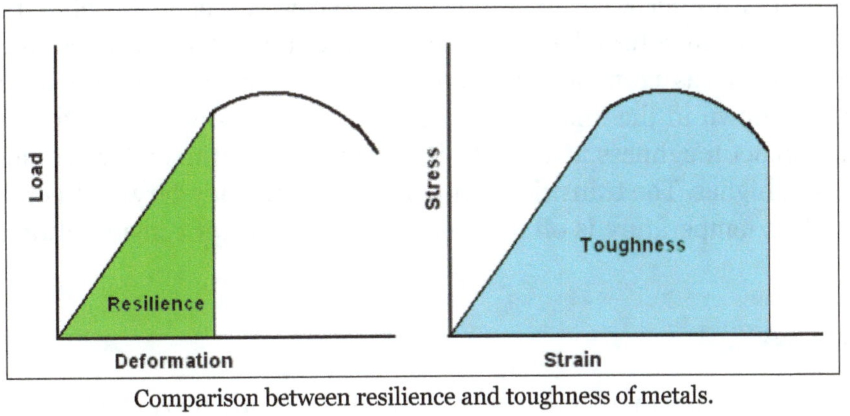

Comparison between resilience and toughness of metals.

Impact Toughness

Three of the toughness properties that will be discussed in more detail are 1) impact toughness, 2) notch toughness and 3) fracture toughness.

The impact toughness (AKA Impact strength) of a material can be determined with a Charpy or Izod test. These tests are named after their inventors and were developed in the early 1900's before fracture mechanics theory was available. Impact properties are not directly used in fracture mechanics calculations, but the economic impact tests continue to be used as a quality control method to assess notch sensitivity and for comparing the relative toughness of engineering materials.

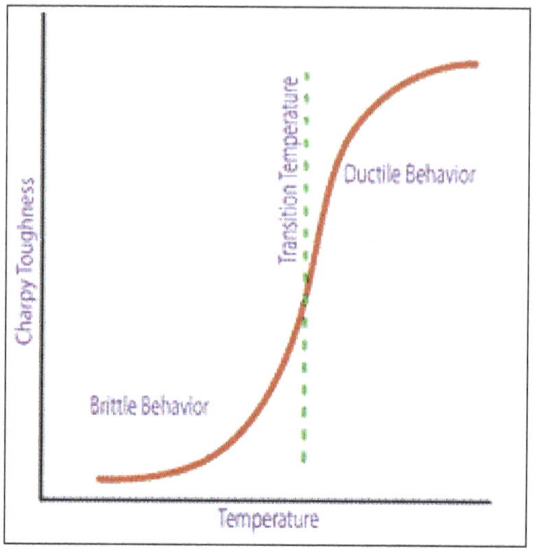

The two tests use different specimens and methods of holding the specimens, but both tests make use of a pendulum-testing machine. For both tests, the specimen is broken by a single overload event due to the impact of the pendulum. A stop pointer is used to record how far the pendulum swings back up after fracturing the specimen. The impact toughness of a metal is determined by measuring the energy absorbed in the fracture of the specimen. This is simply obtained by noting the height at which the pendulum is released and the height to which the pendulum swings after it has struck the specimen. The height of the pendulum times the weight of the pendulum produces the potential energy and the difference in potential energy of the pendulum at the start and the end of the test is equal to the absorbed energy.

Since toughness is greatly affected by temperature, a Charpy or Izod test is often repeated numerous times with each specimen tested at a different temperature. This produces a graph of impact toughness for the material as a function of temperature. The impact toughness versus temperature graph for a steel is shown in the image. It can be seen that at low temperatures the material is more brittle and impact toughness is low. At high temperatures the material is more ductile and impact toughness is higher. The transition temperature is the boundary between brittle and ductile behavior and this temperature is often an extremely important consideration in the selection of a material.

Notch Toughness

Notch toughness is the ability that a material possesses to absorb energy in the presence of a flaw.

As mentioned previously, in the presence of a flaw, such as a notch or crack, a material will likely exhibit a lower level of toughness. When a flaw is present in a material, loading induces a triaxial tension stress state adjacent to the flaw. The material develops plastic strains as the yield stress is exceeded in the region near the crack tip. However, the amount of plastic deformation is restricted by the surrounding material, which remains elastic. When a material is prevented from deforming plastically, it fails in a brittle manner.

Notch-toughness is measured with a variety of specimens such as the Charpy Vnotch impact specimen or the dynamic tear test specimen. As with regular impact testing the tests are often repeated numerous times with specimens tested at a different temperature. With these specimens and by varying the loading speed and the temperature, it is possible to generate curves such as those shown in the graph. Typically only static and impact testing is conducted but it should be recognized that many components in service see intermediate loading rates in the range of the dashed red line.

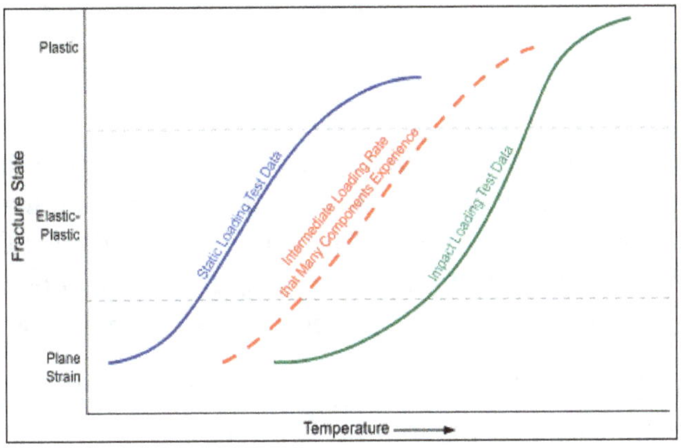

Fracture Toughness

In materials science, fracture toughness is a property which describes the ability of a material containing a crack to resist fracture, and is one of the most important properties of any material for virtually all design applications. It is denoted K_{1c} and has the units of $MPa\sqrt{m}$.

The subscript '1c' denotes mode 1 crack opening or plain strain, the material has to be too thick to shear, mode 2, or tear, mode 3.

Fracture toughness is a quantitative way of expressing a material's resistance to brittle fracture when a crack is present. If a material has a large value of fracture toughness it will probably undergo ductile fracture. Brittle fracture is very characteristic of materials with a low fracture toughness value.

Fracture mechanics, which leads to the concept of fracture toughness, was largely based on the work of A. A. Griffith who, amongst other things, studied the behaviour of cracks in brittle materials.

Fracture toughness is an indication of the amount of stress required to propagate a preexisting flaw. It is a very important material property since the occurrence of flaws is not completely avoidable in the processing, fabrication, or service of a material/component. Flaws may appear as cracks, voids, metallurgical inclusions, weld defects, design discontinuities, or some combination thereof. Since engineers can never be totally sure that a material is flaw free, it is common practice to assume that a flaw of some chosen size will be present in some number of components.

This approach uses the flaw size and features, component geometry, loading conditions and the material property called fracture toughness to evaluate the ability of a component containing a flaw to resist fracture.

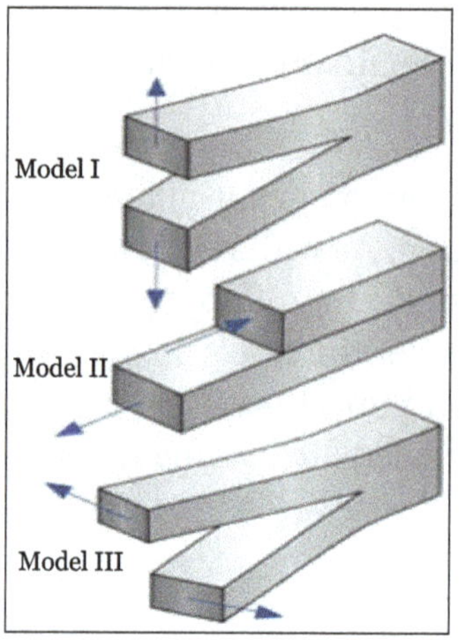

A parameter called the stress-intensity factor (K) is used to determine the fracture toughness of most materials. A Roman numeral subscript indicates the mode of fracture and the three modes of fracture are illustrated in the image to the right. Mode I fracture is the condition in which the crack plane is normal to the direction of largest tensile loading. This is the most commonly encountered mode and, therefore, for the remainder of the material we will consider $K_{I.}$

The stress intensity factor is a function of loading, crack size, and structural geometry. The stress intensity factor may be represented by the following equation:

$$K_I = \sigma\sqrt{\pi a \beta}$$

Where: K_I is the fracture toughness in

σ is the applied stress in MPa or psi.

a is the crack length in meters or inches.

π is a crack length and component geometry.

B factor that is different for each specimen and is Dimensionless.

There are several variables that have a profound influence on the toughness of a material. These variables are:

- Strain rate (rate of loading)
- Temperature
- Notch effect.

A metal may possess satisfactory toughness under static loads but may fail under dynamic loads or impact. As a rule ductility and, therefore, toughness decrease as the rate of loading increases. Temperature is the second variable to have a major influence on its toughness. As temperature is lowered, the ductility and toughness also decrease. The third variable is termed notch effect, has to do with the distribution of stress. A material might display good toughness when the applied stress is uniaxial; but when a multiaxial stress state is produced due to the presence of a notch, the material might not withstand the simultaneous elastic and plastic deformation in the various directions.

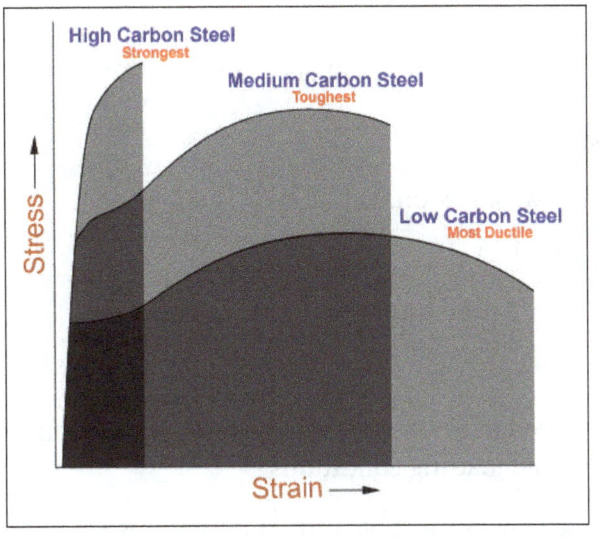

There are several standard types of toughness test that generate data for specific loading conditions and component design approaches.

Hardness

Hardness is the resistance to plastic deformation (e.g., a local dent or scratch). Thus, it is a measure of plastic deformation, as is the tensile strength, so they are well correlated. Historically, it was measured on an empirically scale, determined by the ability of a material to scratch another, diamond being the hardest and talc the softer. Now we use standard tests, where a ball or point is pressed into a material and the size of the dent is measured. There are a few different hardness tests: Rockwell, Brinell, Vickers, etc. They are popular because they are easy and non-destructive (except for the small dent).

Hardness is the resistance of a material to localized deformation. The term can apply to deformation from indentation, scratching, cutting or bending. In metals, ceramics and most polymers, the deformation considered is plastic deformation of the surface. For elastomers and some polymers, hardness is defined at the resistance to elastic deformation of the surface. The lack of a fundamental definition indicates that hardness is not be a basic property of a material, but rather a composite one with contributions from the yield strength, work hardening, true tensile strength, modulus, and others factors. Hardness measurements are widely used for the quality control of materials because they are quick and considered to be nondestructive tests when the marks or indentations produced by the test are in low stress areas.

There are a large variety of methods used for determining the hardness of a substance.

Failure Theories

Theories of failure are those theories which help us to determine the safe dimensions of a machine component when it is subjected to combined stresses due to various loads acting on it during its functionality.

Some examples of such components are as follows:

- I.C. engine crankshaft.
- Shaft used in power transmission.
- Spindle of a screw jaw.
- Bolted and welded joints used under eccentric loading.
- Ceiling fan rod.

Theories of failure are employed in the design of a machine component due to the unavailability of failure stresses under combined loading conditions.

Theories of failure play a key role in establishing the relationship between stresses induced under combined loading conditions and properties obtained from tension test like ultimate tensile strength (S_{ut}) and yield strength (S_{yt}).

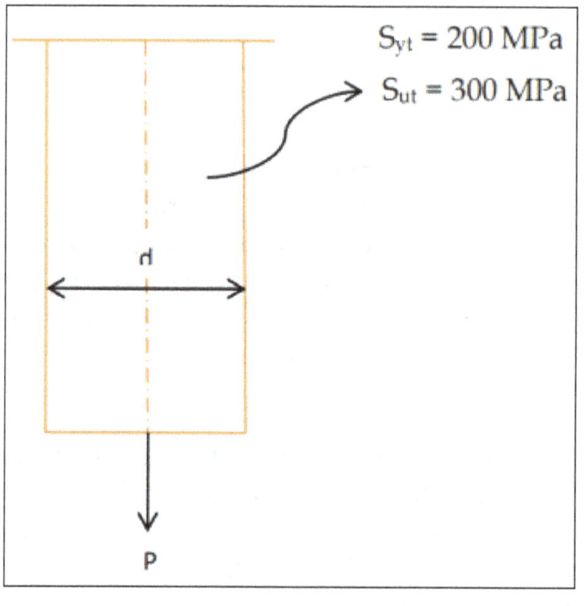

Directly we can get (d) without using any failure theory because only uniaxial load (P),

$$\sigma_1 \leq S_{yt}$$

$$\frac{4P}{\pi d^2} \leq S_{yt}$$

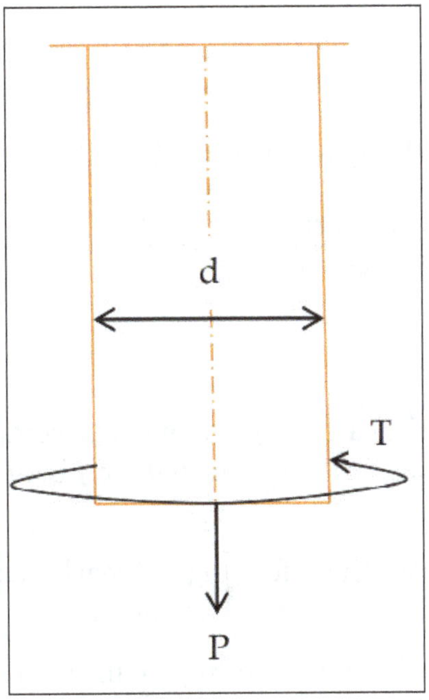

Member is subjected to both Twisting moment and uniaxial load, hence combined loading conditions. We cannot determine (d) directly in this case because failure stresses under combined loading conditions are unknown. So, different scientists give relationships between Stresses induced under combined loading conditions and (S_{yt} and S_{ut}) obtained using tension test which are called theories of failure.

Various Theories of Failure

1. Maximum Principal Stress theory also known as Rankine's Theory.
2. Maximum Shear Stress theory or Guest and Tresca's Theory.
3. Maximum Principal Strain theory also known as St. Venant's Theory.
4. Total Strain Energy theory or Haigh's Theory.
5. Maximum Distortion Energy theory or Vonmises And Hencky's Theory.

Maximum Principal Stress Theory (M.P.S.T)

According to M.P.S. T,

Condition for failure is,

Maximum principal stress (σ_1) failure stresses (S_{yt} or S_{ut})

and Factor of safety (F.O.S) = 1

If σ_1 is +ve then S_{yt} or S_{ut},

σ_1 is –ve then S_{yc} or S_{uc}.

Condition for Safe Design

Factor of safety (F.O.S) > 1

Maximum principal stress (σ_1) ≤ Permissible stress (σ per),

where permissible stress = $\dfrac{\text{Failure stress}}{\text{Factor of safety}} = \dfrac{S_{yt}}{N}$ *or* $\dfrac{S_{ut}}{N}$

$$\sigma_1 \leq \frac{S_{yt}}{N} \text{ or } \frac{S_{ut}}{N}.$$

- This theory is suitable for the safe design of machine components made of brittle materials under all loading conditions (tri-axial, biaxial etc.) because brittle materials are weak in tension.
- This theory is not suitable for the safe design of machine components made of ductile materials because ductile materials are weak in shear.
- This theory can be suitable for the safe design of machine components made of ductile materials under following state of stress conditions.
    - Uniaxial state of stress – Absolute $\tau_{\max} = \dfrac{\sigma_1}{2}$.
    - Biaxial state of stress when principal stresses are like in nature – Absolute $\tau_{\max} = \dfrac{\sigma_1}{2}$.
    - Under hydrostatic stress condition (shear stress in all the planes is zero).

Maximum Shear Stress Theory (M.S.S.T)

Condition for failure,

Maximum shear stress induced at a critical > Yield strength in shear under tensile

point under triaxial combined stress test

$$\text{Absolute } \tau_{\max} > (S_{ys})_{T.T} \text{ or } \frac{S_{yt}}{2}$$

$(S_{ys})_{T.T}$ → unknown therefore use S_{yt}

Condition for Safe Design

Maximum shear stress induced at a critical ≤ Permissible shear stress (τ_{per})

tensile point under triaxial combined stress.

where,

Permissible shear stress = $\dfrac{\text{Yield strength in shear under tension test}}{\text{Factor of safety}} = \dfrac{(S_{ys})_{T.T}}{N} = \dfrac{S_{yt}}{2N}$

Absolute $\tau_{\max} \le \frac{\left(S_{ys}\right)_{T.T}}{N}$ or $\frac{S_{yt}}{2N}$

For tri-axial state of stress,

$$\text{larger of}\left[\left|\frac{\sigma_1-\sigma_2}{2}\right|,\left|\frac{\sigma_2-\sigma_3}{2}\right|,\left|\frac{\sigma_3-\sigma_1}{2}\right|\right] \le \frac{S_{yt}}{2N}$$

$$\text{larger of}\left[|\sigma_1-\sigma_2|,|\sigma_2-\sigma_3|,|\sigma_3-\sigma_1|\right] \le \frac{S_{yt}}{N}$$

For Biaxial state of stress, $\sigma_3 = 0$

$$\left|\frac{\sigma_1}{2}\right| or \left|\frac{\sigma_1-\sigma_2}{2}\right| \le \frac{S_{yt}}{2N}$$

$$|\sigma_1| \le \frac{S_{yt}}{N} \text{ when } \sigma_1, \sigma_2$$

$$|\sigma_1-\sigma_2| \le \frac{S_{yt}}{N} \text{ when } \sigma_1, \sigma_2$$

- M.S.S.T and M.P.S.T will give same results for ductile materials under uniaxial state of stress and biaxial state of stress when principal stresses are like in nature.
- M.S.S.T is not suitable under hydrostatic stress condition.
- This theory is suitable for ductile materials and gives oversafe design i.e. safe and uneconomic design

Maximum Principal Strain Theory (M.P.St.T)

Condition for failure,

Maximum Principal strain (ε_1) Yielding strain under tensile test $(\varepsilon_{Y.P.})_{T.T}$

$$\varepsilon_1 > \left(\varepsilon_{Y.P}\right)_{T.T} \; or \; \frac{S_{yt}}{E}$$

where E is Young's Modulus of Elasticity.

Condition for Safe Design

Maximum Principal strain ≤ Permissible strain,

$$\text{where Permissible strain} = \frac{\text{Yielding strain under tensile test}}{\text{Factor of safety}} = \frac{\left(\varepsilon_{Y.P}\right)_{T.T}}{N} = \frac{S_{yt}}{EN}$$

$$\varepsilon_1 \le \frac{S_{yt}}{EN}$$

$$\frac{1}{E}\left[\sigma_1 - \mu(\sigma_2 + \sigma_3)\right] \le \frac{S_{yt}}{EN}$$

$$\sigma_1 - \mu(\sigma_2 + \sigma_3) \le \frac{S_{yt}}{N}$$

for biaxial state of stress, $\sigma_3 = 0$

$$\sigma_1 - \mu(\sigma_2) \le \frac{S_{yt}}{N}.$$

Total Strain Energy Theory (T.St.E.T)

Condition for failure,

Total Strain Energy per unit volume > Strain energy per unit volume at yield point

(T.S.E. /vol) under tension test (S.E /vol) $_{Y.P.}]_{T.T}$

Condition for Safe Design

Total Strain Energy per unit volume ≤ Strain energy per unit volume at yield point under tension test.

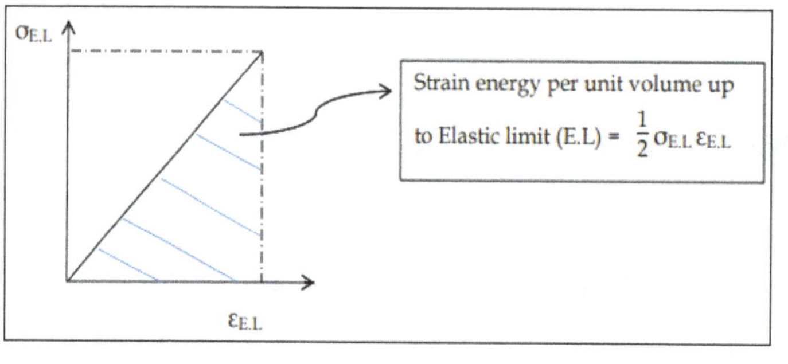

Total Strain Energy per unit volume = $\frac{1}{2}\sigma_1\varepsilon_1 + \frac{1}{2}\sigma_2\varepsilon_2 + \frac{1}{2}\sigma_3\varepsilon_3$

(triaxial)

$$\left.\begin{aligned}
\varepsilon_1 &= \frac{1}{E}\left[\sigma_1 - \mu(\sigma_2 + \sigma_3)\right] \\
\varepsilon_2 &= \frac{1}{E}\left[\sigma_2 - \mu(\sigma_1 + \sigma_3)\right] \\
\varepsilon_3 &= \frac{1}{E}\left[\sigma_3 - \mu(\sigma_1 + \sigma_2)\right]
\end{aligned}\right]$$

By substituting equations $\frac{1}{2}\sigma_1\varepsilon_1 + \frac{1}{2}\sigma_2\varepsilon_2 + \frac{1}{2}\sigma_3\varepsilon_3$ and Total Strain Energy per unit volum $\leq$ Strain energy per unit volume at yield point under tension test.

$$\text{T.S.E. / vol} = \frac{1}{2E}\left[\sigma_1^2 + \sigma_2^2 + \sigma_3^2 - 2\mu\left(\sigma_1\sigma_2 + \sigma_2\sigma_3 + \sigma_3\sigma_1\right)\right]$$

To get $[(\text{S.E /vol})_{\text{Y.P.}}]_{\text{T.T}}$,

Substitute $\sigma_1 = \sigma = \frac{S_{yt}}{N}, \sigma_2 = \sigma_3 = 0$

$$[(\text{S.E /vol})_{\text{Y.P.}}]_{\text{T.T}} = \frac{1}{2E}\left(\frac{S_{yt}}{N}\right)^{2}$$

By Substituting equations $\text{T.S.E. / vol} = \frac{1}{2E}\left[\sigma_1^2 + \sigma_2^2 + \sigma_3^2 - 2\mu\left(\sigma_1\sigma_2 + \sigma_2\sigma_3 + \sigma_3\sigma_1\right)\right]$ and $[(\text{S.E /vol})_{\text{Y.P.}}]_{\text{T.T}} = \frac{1}{2E}\left(\frac{S_{yt}}{N}\right)^{2}$ in equation (Total Strain Energy per unit volume $\leq$ Strain energy per unit volume at yield point under tension test) the following equation is obtained,

$$\sigma_1^2 + \sigma_2^2 + \sigma_3^2 - 2\mu\left(\sigma_1\sigma_2 + \sigma_2\sigma_3 + \sigma_3\sigma_1\right) \leq \left(\frac{S_{yt}}{N}\right)^{2}$$

for biaxial state of stress, $\sigma_3 = 0$

$$\sigma_1^2 + \sigma_2^2 - 2\mu\sigma_1\sigma_2 \leq \left(\frac{S_{yt}}{N}\right)^2$$

1. Eqn $\sigma_1^2 + \sigma_2^2 - 2\mu\sigma_1\sigma_2 \leq \left(\frac{S_{yt}}{N}\right)^2$ is an equation of ellipse ($x^2 + y^2 - xy = a^2$).

2. Semi major axis of the ellipse $= \frac{Syt}{\sqrt{1-\mu}} = \frac{Syt}{\sqrt{0.7}} = 1.0\,S_{yt}$

   Semi minor axis of the ellipse $= \frac{Syt}{\sqrt{1+\mu}} = \frac{Syt}{\sqrt{1.3}} = 0.87\,S_{yt}$

   For $\mu = 0.3$

3. Total strain energy theory is suitable under hydrostatic stress condition.

Maximum Distortion Energy Theory (M.D.E.T)

Condition for Failure

Maximum Distortion Energy/volume (M.D.E/vol) > Distortion energy/volume at yield point under tension test $(\text{D.E/vol}_{\text{Y.P.}}]_{\text{T.T}}$.

Condition for Safe Design

Maximum Distortion Energy/volume ≤ Distortion energy/volume at yield point under tension test

T.S.E/vol = Volumetric S.E/vol + D.E/vol

D.E/vol = T.S.E/vol - Volumetric S.E /vol

Under hydrostatic stress condition, D.E/vol = 0 and

Under pure shear stress condition, Volumetric S.E/vol = 0

From equation,

$$\text{T.S.E / vol} = \frac{1}{2E}\left[\sigma_1^2+\sigma_2^2+\sigma_3^2-2\mu\left(\sigma_1\sigma_2+\sigma_2\sigma_3+\sigma_3\sigma_1\right)\right]$$

Volumetric S.E/vol $= \frac{1}{2}$ (Average stress) (Volumetric strain),

$$= \frac{1}{2}\left(\frac{\sigma_1+\sigma_2+\sigma_3}{3}\right)\left[\left(\frac{1-2\mu}{E}\right)\left(\sigma_1+\sigma_2+\sigma_3\right)\right]$$

$$\text{Vol S.E / vol} = \frac{1-2\mu}{6E}\left(\sigma_1+\sigma_2+\sigma_3\right)^2.$$

From equation D.E/vol = T.S.E/vol - Volumetric S.E /vol and Vol S.E / vol $= \frac{1-2\mu}{6E}\left(\sigma_1+\sigma_2+\sigma_3\right)^2$

$$\text{D.E / vol} = \frac{1+\mu}{6E}\left[\left(\sigma_1-\sigma_2\right)^2+\left(\sigma_2-\sigma_3\right)^2+\left(\sigma_3-\sigma_1\right)^2\right]$$

To get $[(\text{D.E/vol})_{\text{Y.P.}}]_{\text{T.T}}$,

Substitute $\sigma_1=\sigma=\frac{S_{yt}}{N}, \sigma_2=\sigma_3=0$ in equation above $\left[(\text{D.E / vol})_{\text{Y.P.}}\right]_{\text{T.T}} = \frac{1+\mu}{3E}\left(\frac{S_{yt}}{N}\right)^{\wedge}2$

Substituting above equation in the condition for safe design, the following equation is obtained,

$$\left[\left(\sigma_1-\sigma_2\right)^2+\left(\sigma_2-\sigma_3\right)^2+\left(\sigma_3-\sigma_1\right)^2\right] \le 2\left(\frac{S_{yt}}{N}\right)^{\wedge}2.$$

For biaxial state of stress, $\sigma_3 = 0$

$$\sigma_1^2+\sigma_2^2-\sigma_1\sigma_2 \le \left(\frac{S_{yt}}{N}\right)^{\wedge}2$$

1. Equation above is an equation of ellipse.

2. Semi major axis of the ellipse = $\sqrt{2}\,S_{yt}$.

   Semi minor axis of the ellipse = $\sqrt{\frac{2}{3}}S_{yt}$

3. This theory is best theory of failure for ductile material. It gives safe and economic design.
4. This theory is not suitable under hydrostatic stress condition.

Ration of $\frac{S_{YS}}{S_{yt}}$ by using Theories of Failure

1. S_{ys} (Yield strength in shear) is obtained from torsion test.
2. Torsion test is conducted under pure torsion i.e. pure shear state of stress ($\sigma_x = \sigma_y = 0$; $\tau_{xy} = \tau$).
3. Under pure shear state of stress.

$$\sigma_1 = \tau, \sigma_2 = -\tau \text{ and } \tau = \frac{16T}{\pi d^3}$$

4. S_{ys} can also be obtained by applying theories of failure for pure shear state of stress condition.
5. When yielding in shear occurs under pure shear state of stress, $\tau = S_{ys}$.

$\frac{S_{YS}}{S_{yt}}$ in Maximum Principal Stress Theory

According to M.P.S.T,

Considering Factor of safety (N) = 1

$\sigma_1 \le S_{yt}$ or

$\sigma_1 = S_{yt}$

But in pure shear state of stress, $\sigma_1 = \tau$

$\tau = S_{yt}$

When yielding occurs in shear under pure shear state of stress, $\tau = S_{ys}$

$S_{ys} = S_{yt}$

$$\frac{S_{YS}}{S_{yt}} = 1$$

$\frac{S_{YS}}{S_{yt}}$ in Maximum Shear Stress Theory

According to M.S.S.T,

$$|\sigma_1 - \sigma_2| \leq S_{yt}$$

But in pure shear state of stress, $\sigma_1 = \tau$ and $\sigma_2 = -\tau$

$$\tau - (-\tau) = S_{yt}$$
$$2\tau = S_{yt}$$

When yielding occurs in shear under pure shear state of stress, $\tau = S_{ys}$

$$\frac{S_{YS}}{S_{yt}} = \frac{1}{2}.$$

$\frac{S_{YS}}{S_{yt}}$ in Maximum Principal Strain Theory

According to M.P.St.T,

$$\sigma 1 - \mu(\sigma_2) = S_{yt}$$
$$\tau - \mu(-\tau) = S_{yt}$$
$$\tau(1+\mu) = S_{yt}$$

$$S_{ys} = \frac{S_{yt}}{1+\mu}$$

for $\mu = 0.3$

$$\frac{S_{YS}}{S_{yt}} = 0.77.$$

$\frac{S_{YS}}{S_{yt}}$ in Total Strain Energy Theory

According to T.St.E.T,

$$\sigma_1^2 + \sigma_2^2 - 2\mu\,\sigma_1\,\sigma_2 = S_{yt}^{\;2}$$
$$\tau^2 + \tau^2 + 2\,\tau^2 = S_{yt}^{\;2}$$

$$\tau = \frac{Syt}{\sqrt{2(1+\mu)}}$$

$$S_{ys} = \frac{Syt}{\sqrt{2(1+\mu)}}$$

for μ = 0.3

$$\frac{S_{YS}}{S_{yt}} = 0.62.$$

$\frac{S_{YS}}{S_{yt}}$ in Maximum Distortion Energy Theory

According to M.D.E.T,

$$\sigma_1^2 + \sigma_2^2 - \sigma_1\sigma_2 = S_{yt}^{\ 2}$$

$$\tau^2 + \tau^2 + \tau^2 = S_{yt}^{\ 2}$$

$$\tau = \frac{Syt}{\sqrt{3}}$$

$$S_{ys} = \frac{S_{yt}}{\sqrt{3}}$$

$$\frac{S_{YS}}{S_{Yt}} = 0.577.$$

Equivalent Bending Moment (M_e) and Twisting Moment (T_e) Equations

These equations should be used when the component is subjected to both Bending Moment and Twisting Moment simultaneously.

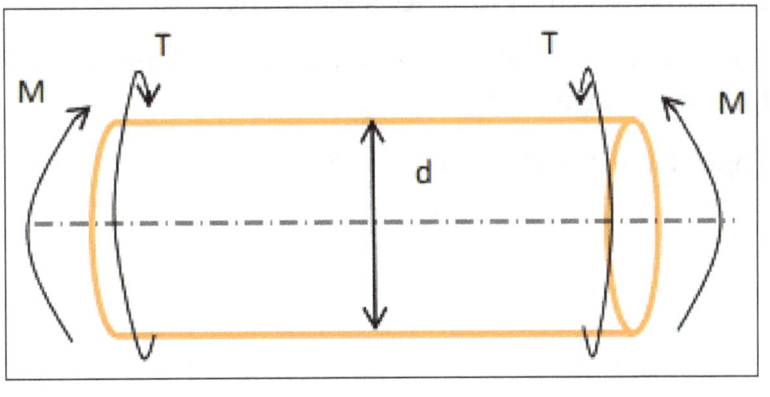

T.O.F	M_e and T_e Equations
M.P.S.T	$M_e = \frac{1}{2}\left[M + \sqrt{M^2 + T^2}\right] = \frac{\pi}{32} d^3 \sigma_{per}$

M.S.S.T	$T_e = \sqrt{M^2 + T^2} = \frac{\pi}{16} d^3 \tau_{per}$
M.D.E.T	$M_e = \sqrt{M^2 + \frac{3}{4} T^2} = \frac{\pi}{32} d^3 \sigma_{per}$

Normal Stress Equations (σ_t Equations)

Normal stress equations should be used when a point in a component is subjected to normal stress in one direction only and a shear stress.

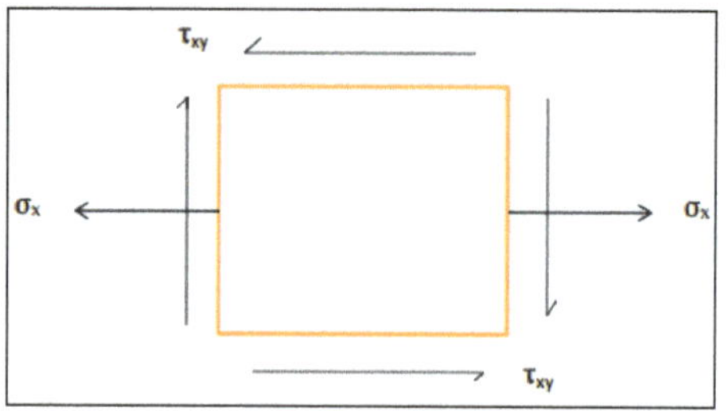

T.O.F	σ_t equations
M.P.S.T	$\sigma_t = \frac{1}{2}\left[\sigma_x + \sqrt{\sigma_x^2 + 4\tau_{xy}{}^2}\right] = \frac{S_{yt}}{N}$
M.S.S.T	$\sigma_t + \sqrt{\sigma_x^2 + 4\tau_{xy}{}^2} = \frac{S_{yt}}{N}$
M.D.E.T	$\sigma_t + \sqrt{\sigma_x^2 + 3\tau_{xy}{}^2} = \frac{S_{yt}}{N}$

Shape of Safe Boundaries for Theories of Failure

Graphical representation or safe boundaries are used to check whether the given dimensions of a component are safe or not under given loading conditions.

As per theories of failure for ductile material, $S_{yc} = - S_{yt}$

(a) M.P.S.T: Square

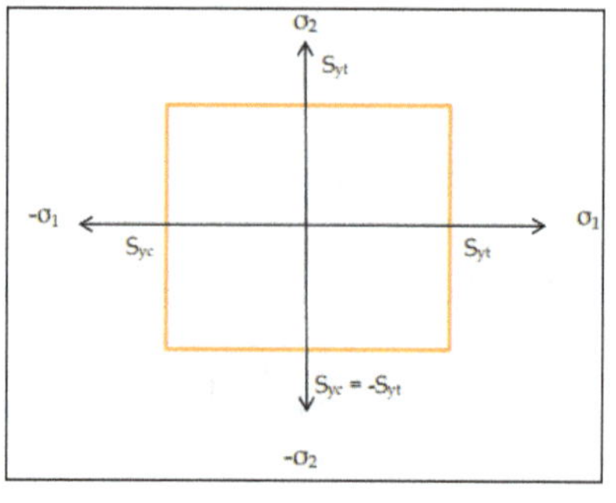

(b) M.S.S.T: Hexagon

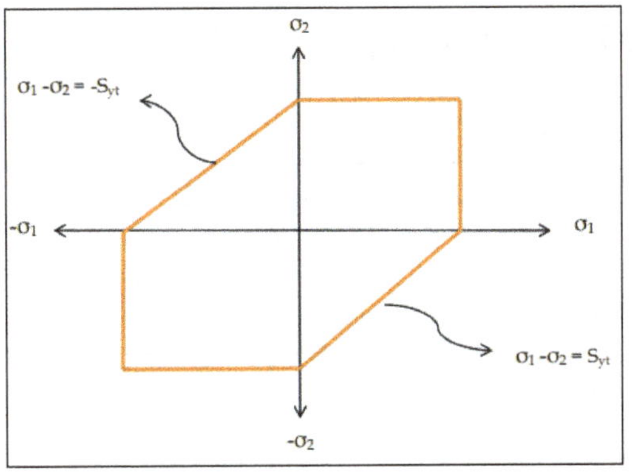

(c) M.P.St.T: Rhombus

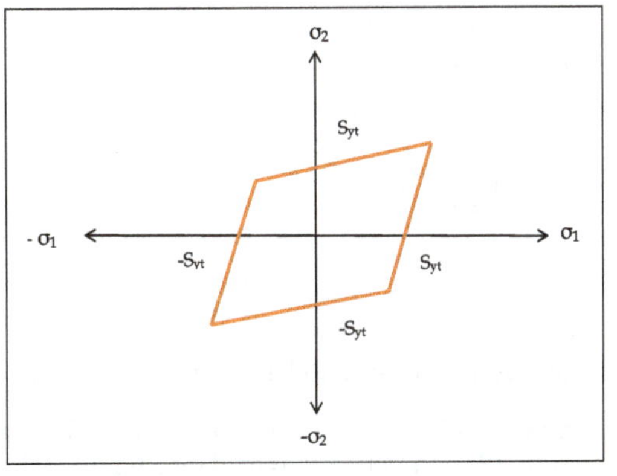

(d) M.D.E.T: Ellipse

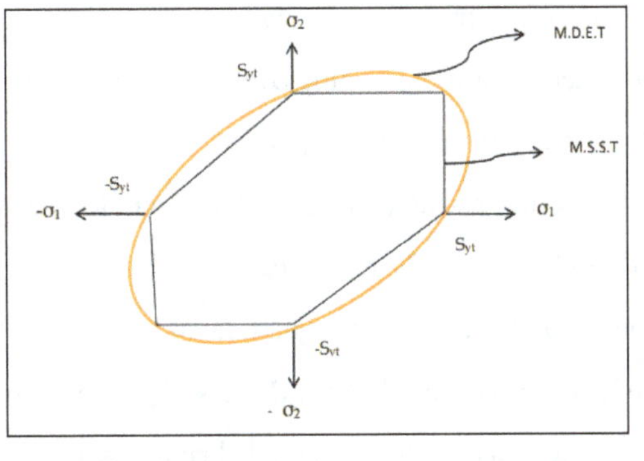

(a) Semi major axis of the ellipse = $\sqrt{2}\ S_{yt}$,

Semi minor axis of the ellipse = $\sqrt{\frac{2}{3}}S_{yt}$.

(b) As the area bounded by the curve increases, failure stresses increases thereby decreases dimensions and hence cost of safety.

In all the quadrants area bounded by the MDET curve > Aread bounded by MSST curve

Hence,

$(\text{Dimensions})_{MDET} < (\text{Dimensions})_{MSST}$.

(c) T.St.E.T: Ellipse,

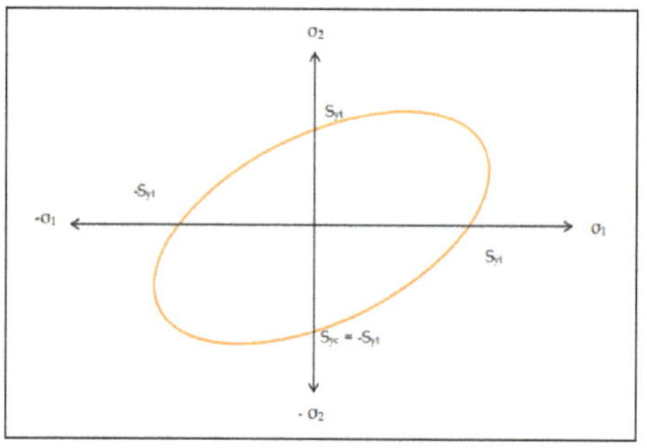

Semi- major axis of the ellipse $= \dfrac{Syt}{\sqrt{1-\mu}}$

Semi- minor axis of the ellipse $= \dfrac{Syt}{\sqrt{1+\mu}}$.

References

- StressStrain, mechanical, materials: nde-ed.org, Retrieved 14 June, 2019
- Rees, David (2006). Basic Engineering Plasticity: An Introduction with Engineering and Manufacturing Applications. Butterworth-Heinemann. ISBN 0-7506-8025-3. Archived from the original on 2017-12-22
- What-is-the-stress-strain-curve: scienceabc.com, Retrieved 4 April, 2019
- Poissons-ratio, physics: byjus.com, Retrieved 18 January, 2019
- Elasticity-stress-strain-and-fracture: lumenlearning.com, Retrieved 7 March, 2019
- Hookes-law: lumenlearning.com, Retrieved 13 July, 2019
- Bulk-modulus-definition-and-examples-4175476: thoughtco.com, Retrieved 9 May, 2019
- Shear-modulus-4176406: thoughtco.com, Retrieved 12 June, 2019
- Plasticity, science: britannica.com, Retrieved 20 August, 2019
- Plasticity-Introduction: engineering.auckland.ac.nz, Retrieved 30 April, 2019
- Theories-of-failure: thegateacademy.com, Retrieved 11 February, 2019

4

Fluid Mechanics and Hydraulic Machines

Fluid mechanics is the domain of physics which studies the mechanics of different fluids such as liquids, plasma and gases, as well as the forces which act on them. There are a number of hydraulic machines like hydraulic turbines and pumps which are designed using fluid mechanics. The topics elaborated in this chapter will help in gaining a better perspective about the diverse concepts related to fluid mechanics and hydraulic machines.

Fluid Mechanics

A fluid is a defined as substance that can flow, that doesn't maintain a fixed shape. Gases and liquids are usually considered fluids. Any object, whether a solid, a gas, a liquid, or a plasma (a collection of ionized particles), has a density. The density is defined as the mass per unit volume and is given the Greek symbol rho (ρ). So the density is defined as,

$$\rho = \mathrm{m}/\mathrm{V}.\ \left(\mathrm{kg}/\mathrm{m}^3\right)$$

A substance which is more dense will have more mass for a given volume. If we hold a cubic centimeter of lead, it will be heavier than a cubic centimeter of wood because lead has a higher density than wood.

Another way of specifying the density of an object is to compare its density to the density of water at 4.0 °C, (which is 1.00×10^3 kg/m³). The specific gravity is defined as the ratio of the density of a substance to the density of water at 4.0 °C. Lead, with a specific gravity of 11.3, has a density of 11.3×10^3 kg/m³.

Pressure

Pressure is defined as the force per unit area. P = F/A. The force acts perpendicular to the surface area A. If a person stands on a wooden floor with spiked metal cleats, the floor will be damaged because the person's weight is not spread out over a very large area. However, if the person wears normal shoes, the weight is spread out over a larger area, so the pressure is smaller, and the floor will not be damaged. Snow shoes spread the weight out over an even larger area, and so the pressure on the snow is less, and a person does not sink into the snow. A person can lay on a bed of nails because there are so many nails that the person's weight is still spread out over a lot of area, and the pressure is not very high.

The SI units of pressure is given by N/m² which is given the name a Pascal (Pa), named after the French philosopher, theologian, and scientist Blaise Pascal.

$$1\ \mathrm{N}/\mathrm{m}^2 = 1\ \mathrm{Pa}.$$

In the U.S. we also use lb/in^2 or psi (pounds per square inch).

Because the atmosphere is always pushing down on us with a certain pressure, another common unit of measurement, often used when discussing fluids, is called an atmosphere. This is defined as the standard pressure that the atmosphere presses on us at sea level and is equal to 1.013×10^5 Pa. This is given the pressure unit of atmospheres.

Finally, we will see that other fluids can push on us just like the air pushes on us. A commonly used measure of pressure comes from the fact that if we were immersed in a column of mercury 760 mm high, that would exert the same pressure on us as the atmosphere. We give this unit of mm/Hg (millimeters of mercury) a special name called a torr. So,

1 atmosphere = 760 torr (mm/Hg) = 1.013×10^5 Pa

A water bed is 2.00 m square and 30.0 cm deep. (a) What is its weight? (b) What pressure does the bed exert on the floor, assuming the entire lower surface contacts the floor? (Volume = length × width × height, Area = length × width).

Properties of Pressure in a Fluid

There are three important observations about pressure in a fluid:

1) At any point that a fluid is in contact with a surface, the pressure is directed perpendicular to the surface.

2) At any point inside a fluid, the pressure is directed in all directions with the same magnitude. For the block in this figure, the pressure is not exactly the same because the different faces are at different depths. But if the block is infinitesimally small, then the pressure in all directions is exactly the same.

3) The pressure at any point in a fluid depends only on the depth of the point. Suppose there is a volume of fluid with a uniform density which has a depth h and area A.

What forces act on this volume of fluid? There is the force from the atmosphere above the liquid $F_0 = P_0A$. There is the weight of the liquid Mg. There is also the force of the liquid pushing up on the column of liquid at depth h.

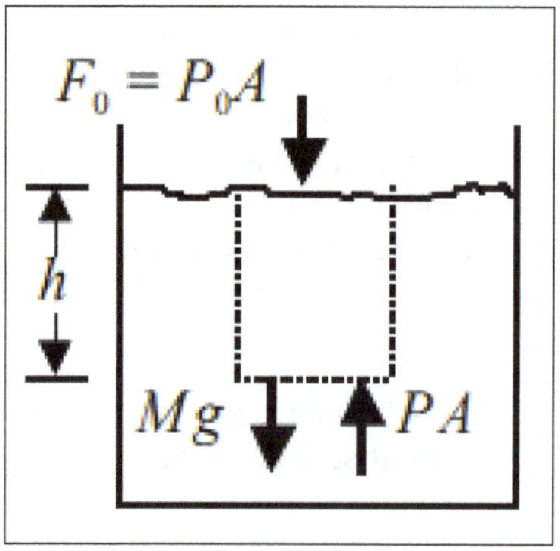

Since there is no acceleration of the liquid at the depth h, the forces must balance each other.

$$\Sigma F_y = 0$$
$$PA - Mg - P_0A = 0$$
$$PA - \rho Vg - P_0A = 0$$
$$PA - \rho hAg - P_0A = 0$$

$$P = \rho gh + P_0$$

So the pressure at any depth h in a fluid is equal to the pressure outside of the fluid (P_0) plus the fluid pressure (ρhg).

For many circumstances, the pressure outside of the fluid is the pressure of the earth's atmosphere at that point. For instance, the pressure in your tires is really the pressure of the air in your tires plus the pressure of the atmosphere on the tires. However, the pressure that is read on a pressure gauge is the pressure which is greater than the atmospheric pressure. It is called gauge pressure. In general, the equation above becomes,

$$P = P_G + P_A$$

where P_G is the gauge pressure and P_A is the atmospheric pressure. We are often most concerned with the gauge pressure or fluid pressure, and not the outside pressure. For fluids, then, we often simply write the fluid pressure, or

$$P_{fluid} = \rho gh$$

A rectangular shaped dam is 70 m high and 180 m wide and water is filled to the top of the dam. (a) What is the pressure on the dam from the water at the top? (b) What is the pressure on the dam from the water at the bottom? (c) What is the average pressure on the dam from the water? (d) What is the total force on the dam? (e) What if we include the pressure from the atmosphere? (f) Suppose that the dam was holding up a thin column of water instead of a huge lake. What would be the pressure on the dam then?

Pascal's Principle

Note that the pressure is only dependent on the depth, and not on the "area" of the column of fluid. This means that the pressure at any depth of a fluid is always the same for all points at that depth. The pressure is not affected by the shape of the vessel. The pressure under 5 meters of water in a thin tube is the same as the pressure under 5 meters of water in a big swimming pool. This is why water always seeks a constant level. The pressure at any uniform height in a single liquid is the same throughout the entire liquid.

In addition, if the outside or external pressure increases, the total pressure increases by the same amount. Pascal realized that this led to an important general principle. Pascal's principle states that pressure applied to a confined fluid increases the pressure throughout the fluid by the same amount. This is how a hydraulic lift works. Fluid is enclosed in a pipe with a small area at one end and a large area at the other. Pressure is applied at the end with the small area. That same pressure is transferred to the end with the large area.

$$P_1 = F_1 / A_1 = P_2 = F_2 / A_2$$
$$F_2 = F_1 (A_2 / A_1)$$

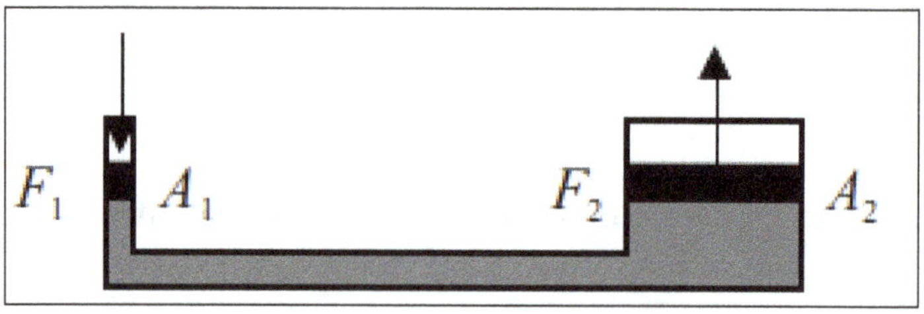

If the area A_2 is much larger than the area A_1, then a small force F_1 can be applied to create a large force F_2 at the output end. This large force can be used to jack up a car or lift heavy objects.

In the hydraulic press used in a trash compactor, the radii of the input piston and the output plunger are 6.4×10^{-3} m and 5.1×10^{-2} m, respectively. What force is applied to the trash when the input force is 330 N?

Measurements of Pressure

Pressure is measured using the two principles, that P = ρgh and Pascal's Principle. If a fluid, like mercury or water is put in a container which is open to the atmosphere at one end and closed at the other, with the closed end having zero pressure (it is a vacuum), then the following situation occurs.

The pressure at any height is equal, so the pressure of the atmosphere, just equals the pressure of the liquid or ρgh.

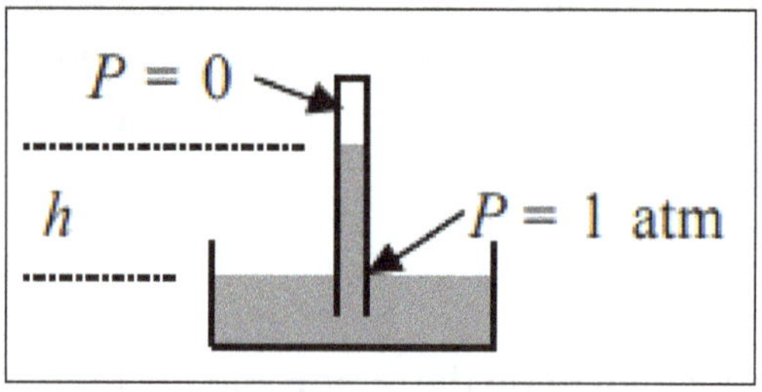

For different liquids with different densities, the height of the column at sea level will be different. For mercury it is 760 mm. For water, it is 10.3 m.

What happens when you suck on a straw? You take the air out of the straw and the atmospheric pressure pushes the liquid up the straw. You do not suck the liquid up. Instead you create a vacuum and the atmosphere pushes the liquid up. Similarly, a water pump pumping at the top of a well can only pump water up 10.3 m, since it works the same way, and atmospheric pressure is 10.3 m of water. To pump water up larger distances, you must use a series of pumps, or place the pump at the bottom of the well, and then it pushes the water up, rather than using atmospheric pressure to push it up.

Buoyancy

Any object which is partially or totally submerged in a liquid has buoyant force acting on it which pushes the object up. That is why a rock appears to weigh less when it is submerged in liquid, or why it is very difficult to push a beach ball underwater. The famous Greek mathematician, Archimedes' developed a principle which describes this around 250 B.C. Archimedes' principle can be stated as anybody completely or partially submerged in a fluid is buoyed up by a force equal to the weight of the fluid displaced by the body.

$$B = w_{fluid}$$

where B is the magnitude of the buoyant force and W_{fluid} is the weight of the displaced fluid. The reason for this is that the pressure of the fluid is dependent on the depth of the fluid. So the pressure at the top of an object is less than the pressure at the bottom of the object which creates a net force.

$$B = P_2A - P_1A$$
$$= (P_2 - P_1)A = \rho ghA$$
$$= \rho gV = mg = w$$

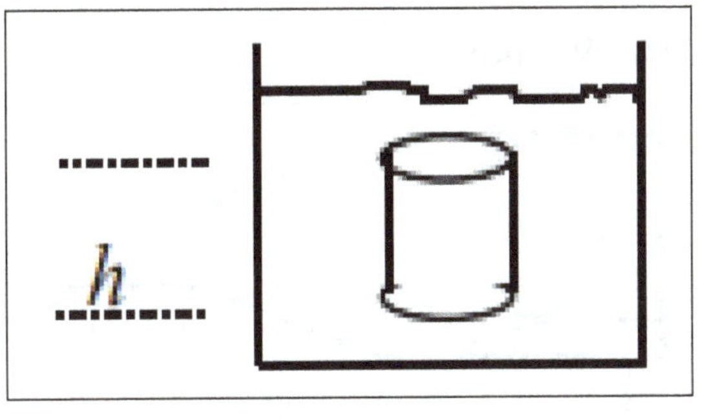

(a) What is the buoyant force on a balloon filled with 1.0 m³ helium at sea level. (b) What is the gravitational force on the same balloon? So the upward (buoyant) force is much greater than the downward force (weight) and the helium balloon rises.

Every fluid exerts a buoyant force. Even the air exerts a buoyant force on your body. Estimate how strong that force is. Archimedes supposedly determined whether or not the king's crown was made of gold by using this law. Suppose a crown is weighed on a scale and it weighs 9.8 N. The crown is then weighed in water and it weighs 9.00 N. Is the crown pure gold?

A raft is made of wood having a density of 600 kg/m³. Its surface area is 5.7 m², and its volume is 0.60 m³. How much of it is below water level?

A hollow log is used as a canoe. It has a length of 3.00 m and a radius of 0.350 m. The canoe weighs 1.00×10^3 N. What is the maximum weight it can carry without sinking?

As the weight in the canoe is increased, it sinks lower in the water. The maximum weight it can carry is when the water level is at the top of the canoe's side. The canoe has an upward buoyant force F_B, and two downward forces, the weight of the canoe w_C, and the weight of the cargo, w.

Motion of Fluids

There are many categories of fluids in motion, categorized by whether the fluid flow is steady, or not steady, compressible or incompressible, viscous or nonviscous. In steady flow, the velocity of the fluid particles at any point is constant as time goes by. Different parts of the fluid may be flowing at different rates, but the fluid in one location is always flowing at the same rate. Anincompressible flow is the flow of a fluid which cannot be compressed. Most liquids are nearly incompressible. A viscous fluid is one which does not flow easily, like honey, while a nonviscous fluid is one which flows more easily, like water. We will mostly be concerned with the steady flow of incompressible, nonviscous fluids.

If the flow is steady, then the velocity of the fluid particles at any point is a constant with time. The various layers of the fluid slide smoothly past each other. This is called streamline or laminar flow. Above some certain velocity, the flow is not smooth and becomes turbulent. Illustrations of turbulent and laminar flow are shown in Figure.

We first consider the steady flow of a fluid through an enclosed pipe. We want to determine how the speed of the fluid changes when the size of the pipe changes.

At some point along the pipe, we look at how much fluid flows past us (Δm) in a short period of time (Δt).

$$\Delta m/\Delta t = \rho \Delta V/\Delta t = \rho A \Delta l/\Delta t = \rho A v.$$

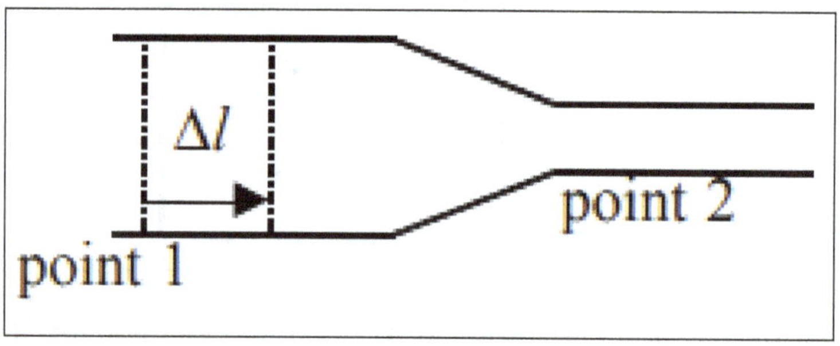

This is the flow rate for any point along the pipe. Because no fluid flows in or out of the sides, the mass flowing past any point during a short period of time must be the same as the mass flowing past any other point, so

$$\rho_1 A_1 v_1 = \rho_2 A_2 v_2$$

where the subscript 1 and 2 refer to two different points along the pipe. This equation is called the equation of continuity. If the fluid is incompressible, then the density is the same at all points along the pipe and this equation becomes

$$A_1 v_1 = A_2 v_2$$

We see that if the cross sectional area is decreased, then the flow rate increases. This is demonstrated when you hold your finger over part of the outlet of a garden hose. Because you decrease the cross sectional area, the water velocity increases.

The flow rate or flux is defined as the rate with which the fluid flows and is given by the variable Q.

$$Q = \Delta V/\Delta t = Av$$

A water hose with a radius of 1.00 cm is used to fill a 20.0 liter bucket. If it takes 1.00 min to fill the bucket, what is the speed, v, at which the water leaves the hose? (1.00 liters = 10^3 cm^3).

The approximate inside diameter of the aorta is 1.0 cm (radius = 5.0×10^{-3} m) and that of a capillary is 10 μm = 10×10^{-6} m. The approximate blood flow speed is .3 m/s in the aorta and 0.5 mm/s in the capillaries. Approximately how many capillaries get blood from the aorta?

Bernoulli's Equation

In the 18th century, the Swiss physicists Daniel Bernoulli derived a relationship between the velocity of a fluid and the pressure it exerts. Qualitatively, Bernoulli's principle states that swiftly moving fluids exert less pressure than slowly moving fluids.

Bernoulli's principle is extremely important in our everyday life. It is the primary principle which leads to lift on an airplane wing and allows the plane to fly. It is the primary reason a sailboat can sail into the wind. It is the primary reason a baseball can curve. It is an important reason that smoke is drawn up a chimney.

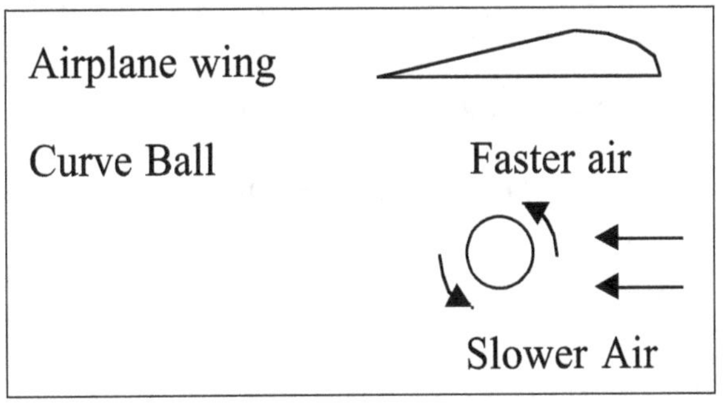

Bernoulli's equation is really a consequence of a fundamental principle of physics: the conservation of energy. It can be derived using energy principles. Consider a fluid moving through a pipe. The pipe's cross sectional area changes, and the pipe changes elevation. At one point the pipe has a cross sectional area of A_1, a height of y_1, a pressure of P_1, a velocity of v_1 and moves a distance of Δx_1 in a time of Δt. At another point P_1 along the pipe these quantities are given by A_2, y_2, P_2, v_2, and Δx_2.

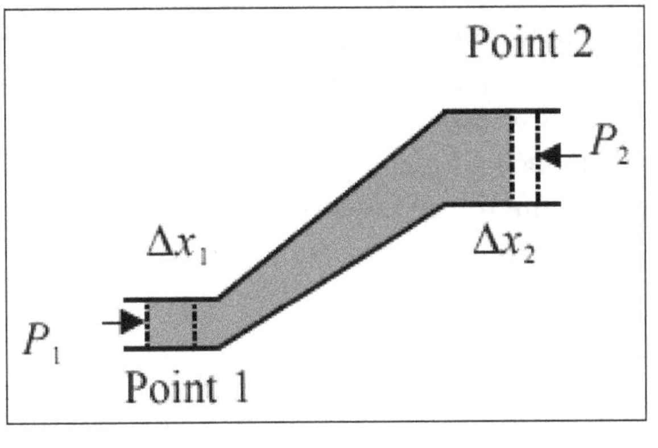

We are going to push a certain amount of fluid up the pipe from point 1 to point 2. P_1 is opposite in direction from P_2 because the rest of the fluid pushes to the left of fluid at point 2 and to the right of the fluid at point 1.

Recall, the fundamental equation that,

$$W_{NC} = \Delta K + \Delta U$$

We want to look at each of these terms individually to derive Bernoulli's equation.

$$\Delta K = (1/2)mv_2^2 - (1/2)mv_1^2 = (1/2)\rho V_2 v_2^2 - (1/2)\rho V_1 v_1^2$$
$$\Delta U = mgy_2 - mgy_1 = \rho V_2 g y_2 - \rho V_1 g y_1$$

$$W_1 = F\Delta x_1 = P_1 A_1 \Delta x_1$$
$$W_2 = F\Delta x_2 = P_2 A_2 \Delta x_2$$

$$W_{NC} = W_1 + W_2 = P_1 A_1 \Delta x_1 - P_2 A_2 \Delta x_2 = P_1 V_1 - P_2 V_2$$

$$W_{NC} = \Delta K + \Delta U$$

$$P_1 V_1 - P_2 V_2 = (1/2)\rho V_2 v_2^2 - (1/2)\rho V_1 v_1^2 + \rho V_2 g y_2 - \rho V_1 g y_1$$

We know that since this is an incompressible fluid, $V_1 = V_2$,

$$P_1 - P_2 = (1/2)\rho v_2^2 - (1/2)\rho v_1^2 + \rho g y_2 - \rho g y_1$$

$$P_1 + (1/2)\rho v_1^2 + \rho g y_1 = P_2 + (1/2)\rho v_2^2 + \rho g y_2$$

which is Bernoulli's equation.

Remember, that Bernoulli's principle applies when a fluid is moving. The pressure is not the same if the velocities are different. Pascal's principle applies when the fluid is stationary. A pressure applied at one point is transferred to every point of the fluid. Pascal's principle can be used for something moving slowly or something that moves a little then stops like a hydraulic jack.

As shown in the book, for a large storage tank of height h which has a small pipe open at the bottom of it, the pressure at the top and at the small opening is atmospheric pressure. This is actually an important point for solving problems. The pressure of any part of a container open to the atmosphere is the same pressure as the atmosphere for the entire section of container that is at the same height as the opening and has the same cross sectional area as the opening. Also in this case, the velocity at the top of the storage tank is basically zero because it has such a huge cross sectional area, so Bernoulli's equation gives the velocity at the bottom of the tank to be:

$$P_1 + (1/2)\rho v_1^2 + \rho g y_1 = P_2 + (1/2)\rho v_2^2 + \rho g y_2$$

$$P_1 = P_2 = 1 \text{ atm}, v_1 = 0, y_1 - y_2 = h, so$$

$$v_2 = \sqrt{2gh}$$

A water pipe is inclined 30° below the horizontal. The radius of the pipe at the upper end is 2.00 cm. If the gauge pressure at a point at the upper end is 0.100 atm, what is the gauge pressure at a point 3.00 m downstream where the pipe has narrowed to 1.00 cm radius? The flow rate is 20.0π cm^3/s.

A hypodermic syringe contains a fluid with the density of water. The barrel of the syringe has a cross sectional area of 2.50×10^{-5} m2 and the cross sectional area of the needle is 2.50×10^{-8} m^2. Before the plunger is pushed, the pressure everywhere is at 1.00 atmospheres. A force, F, of 2.00 N is exerted on the plunger. Assume that the gauge pressure in the needle remains at 1.00 atm. If the syringe is horizontal, what is the speed of the liquid as it flows through the needle and into the arm?

Don't confuse this situation with one where there is a piston pushing on the fluid, but the fluid doesn't move. If you are considering the situation before the fluid starts to move Pascal's principle applies. After the fluid starts to move Bernoulli's principle applies.

Viscosity

Viscosity is the amount of internal friction in a fluid. As stated before, a fluid like honey is very viscous, and a fluid like water has a low viscosity. The viscosity of a fluid is quantitatively given by its coefficient of viscosityη. It takes a constant force to move a fluid, because there is a frictional viscous force opposing the motion. (Remember from Newton's 2nd law that if there was no frictional force, it would not take any external force to keep a body moving at constant velocity). The force it takes to keep a liquid moving in a tube is found to follow the equation,

$$F = \eta A v / l$$

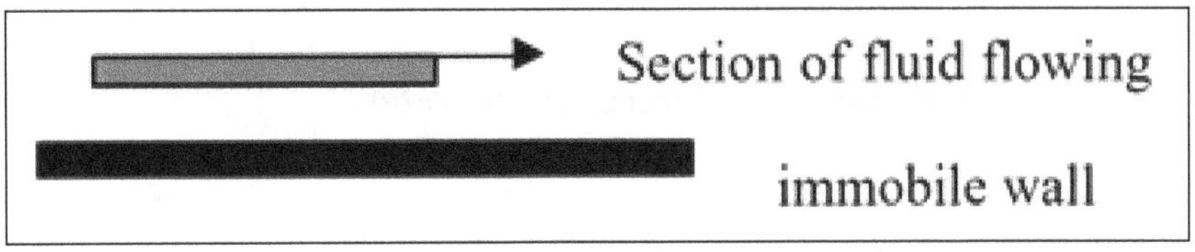

where, A is the area of the fluid in contact with the walls, v is the velocity of fluid, and l is the perpendicular distance from the fluid to the immobile surface.

So, a fluid has more drag near the surface of a stationary plate, and as l approaches zero it takes an infinite force to move the fluid. In other words, the fluid touching the wall does not actually move at all.

Surface Tension and Capillarity

We now deal with what happens at the surface of a fluid. The surface of a fluid does not act like the rest of the fluid, but instead acts something like a stretched membrane. This allows insects to walk on water, and water drops to hang from a faucet. This phenomena is due to attractive forces between the fluid molecules called surface tension. The surface tension is defined as the force per unit length that acts along any line of a surface. Surface tension is defined by,

$$\gamma = F/L$$

where γ is the Greek letter gamma, and is defined as the surface tension. When a small object is placed in a pool of water it depresses the water and this surface tension force tends to hold up the object. Look at the forces on the object.

The vertical forces are given by,

$$\Sigma F = 0$$

$$w - 2\gamma L\cos\theta = 0$$

If the object is totally submerged, then L for each side is half of the circumference, or $L = \pi r$.

Then, $w = 2\gamma\pi r \cos\theta$.

So for a spherical object to float on a liquid, its weight must be approximately less than $2\gamma\pi r$, since the maximum value of $\cos\theta$ is 1. If the weight is more than that, then surface tension will not play a real role in the problem, and we just use Archimedes' principle and buoyancy to analyze the situation.

Surface tension also explains why water rises up slightly in a glass, or can even climb up a very thin tube. This phenomena is called capillarity. Suppose we put a thin tube vertically into a pool of liquid.

Using Newton's second law, we look at the forces on the water,

$$\gamma L - mg = 0$$
$$\gamma(2\pi r) - \rho(\pi r^2 h)g = 0$$

$$h = 2\gamma/(\rho g r)$$

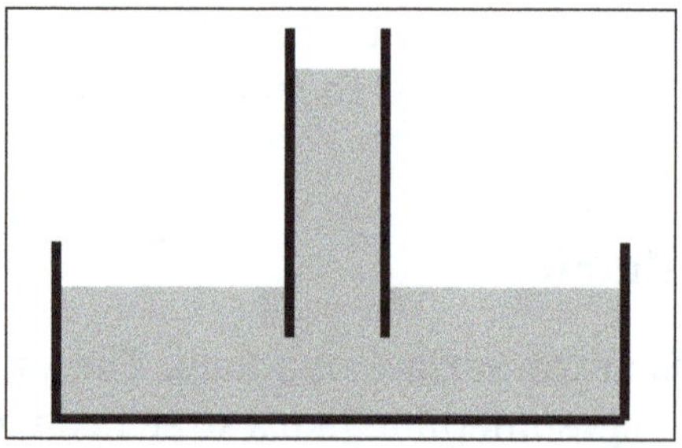

If r gets very large, then h is very small. So the fluid will only rise any appreciable height if the radius of the tube is very small.

Drag

When an object moves through a fluid it experiences a drag. When we move through the ocean of air that we live in, we experience a drag that we call air resistance. This drag, analogous to air resistance, is experienced by any object moving through any fluid. The drag that an object experiences is characterized by something we call the Reynolds number. The Reynold's number is defined as,

$$R = vL\rho/\eta$$

where v is the velocity of the object relative to the fluid, ρ is the density, η is the coefficient of viscosity, and L is a characteristic length of the object which depends on its shape. As long as the Reynold's number is less than about one, then the fluid flow around the object is laminar and the force of drag is proportional to the velocity,

$$F_D \propto v = kv.$$

That means that as long as the Reynold's number is below one, when your speed doubles, the force of drag acting against your direction of motion also doubles. When the Reynold's number is greater than one, the flow of air is no longer laminar. Instead it is turbulent, and the drag is proportional to the square of the velocity,

$$F_D \propto v2$$

This means when you velocity doubles, your drag is increased by a factor of four. The more aerodynamic an object is the lower the Reynold's number. We try to make objects aerodynamic not only to decrease wind resistance, but because if the Reynold's number is below about 1, then the drag of an object will increase linearly with velocity rather than quadratically. On a bicycle, the drag increases approximately with the square of the velocity, so when your speed doubles, the drag from wind resistance is increased by a factor of four.

Let us consider the case of Reynold's number below one with the fluid flow laminar. Then $F_D = kv$, where k depends on the size and shape of the object and on the viscosity of the fluid. For a spherical object, $k = 6\pi r\eta$, so the force of drag becomes $F_D = 6\pi r\eta v$ which is called Stoke's Equation.

Consider an object falling in a fluid. There are three forces acting on it, gravity, buoyancy, and drag. By Newton's second law,

$$mg - B - F_D = ma$$

So when the flow is laminar,

$$mg - B - kv = ma$$

$$\rho_O Vg - \rho_F Vg - kv = ma$$

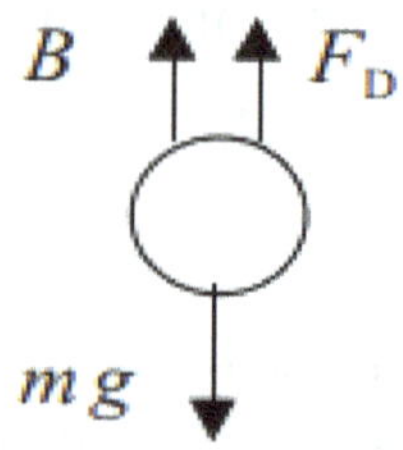

where ρ_O is the density of the object, and ρ_F is the density of the fluid. When the sum of the forces equal, and there is no further acceleration, then the object reaches its terminal velocity. (Remember this is all for the case of laminar flow),

$$\rho_O Vg - \rho_F Vg - kv_T = 0$$

$$v_T = (P_O - P_F)VG / K.$$

Hydraulic Turbines

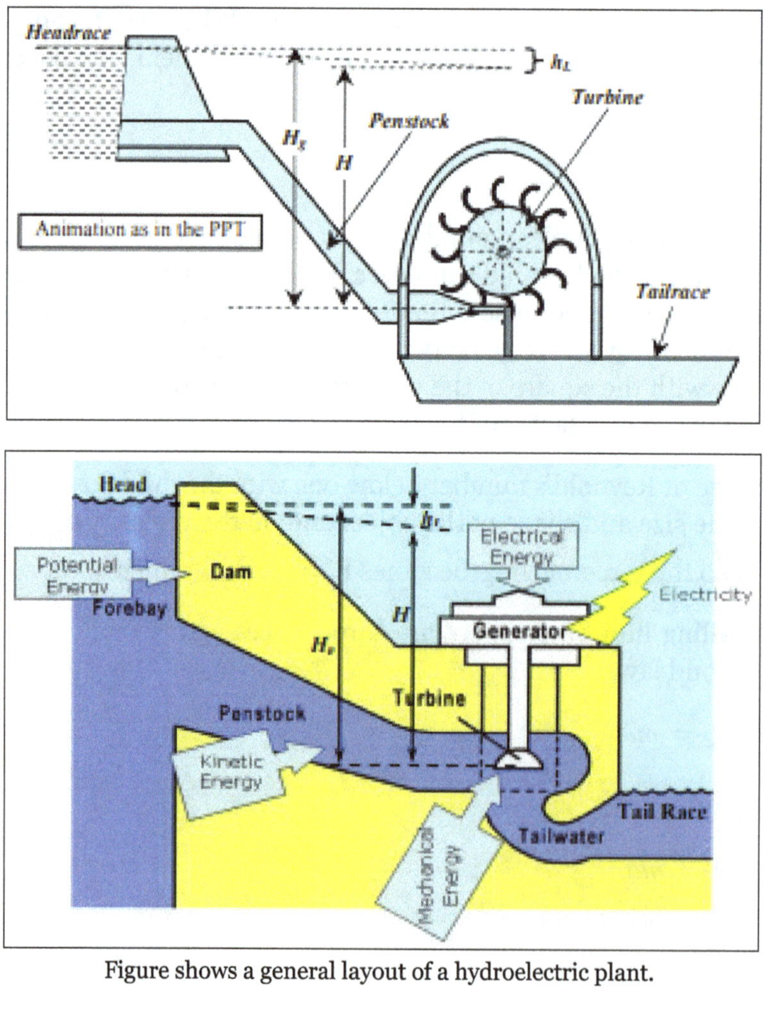

Figure shows a general layout of a hydroelectric plant.

The device which converts hydraulic energy into mechanical energy or vice versa is known as Hydraulic Machines. The hydraulic machines which convert hydraulic energy into mechanical energy are known as Turbines and that convert mechanical energy into hydraulic energy is known as Pumps.

It consists of the following:

- A Dam constructed across a river or a channel to store water. The reservoir is also known as Headrace.
- Pipes of large diameter called Penstocks which carry water under pressure from storage reservoir to the turbines. These pipes are usually made of steel or reinforced concrete.
- Turbines having different types of vanes or buckets or blades mounted on a wheel called runner.
- Tailrace which is a channel carrying water away from the turbine after the water has worked on the turbines. The water surface in the tailrace is also referred to as tailrace.

Important Terms:

- Gross Head (H_g): It is the vertical difference between headrace and tailrace.
- Net Head (H): Net head or effective head is the actual head available at the inlet of the to work on the turbine.

  $H = H_g - hL$

Where h_L is the total head loss during the transit of water from the headrace to tailrace which is mainly head loss due to friction, and is given by,

$$h_f = \frac{4fLV^2}{2gd}$$

Where f is the coefficient of friction of penstock depending on the type of material of penstock.

L is the total length of penstock,

V is the mean flow velocity of water through the penstock,

D is the diameter of penstock and

g is the acceleration due to gravity.

Types of Efficiencies

Depending on the considerations of input and output, the efficiencies can be classified as

- Hydraulic Efficiency
- Mechanical Efficiency

- Overall efficiency.

(i) Hydraulic Efficiency: (ηh)

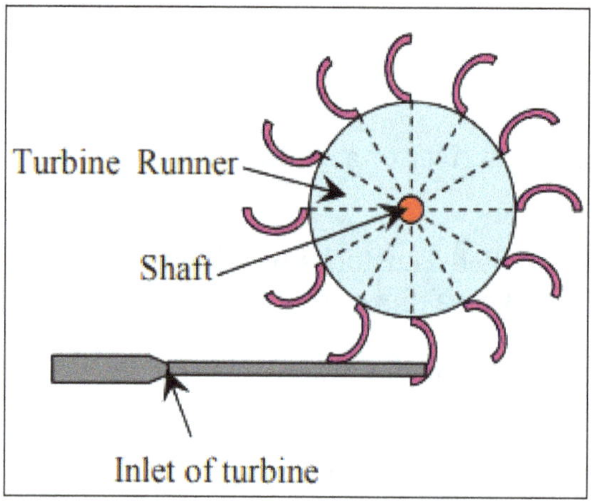

It is the ratio of the power developed by the runner of a turbine to the power supplied at the inlet of a turbine. Since the power supplied is hydraulic, and the probable loss is between the striking jet and vane it is rightly called hydraulic efficiency. If R.P. is the Runner Power and W.P. is the Water Power,

$$\eta h = \frac{R.P.}{W.P.}$$

(ii) Mechanical Efficiency: (η m)

It is the ratio of the power available at the shaft to the power developed by the runner of a turbine. This depends on the slips and other mechanical problems that will create a loss of energy between the runner in the annular area between the nozzle and spear, the amount of water reduces as the spear is pushed forward and vice-versa and shaft which is purely mechanical and hence mechanical efficiency.

If S.P. is the Shaft Power

$$\eta_m = \frac{S.P.}{R.P.}$$

(iii) Overall Efficiency: (η)

It is the ratio of the power available at the shaft to the power supplied at the inlet of a turbine. As this covers overall problems of losses in energy, it is known as overall efficiency. This depends on both the hydraulic losses and the slips and other mechanical problems that will create a loss of energy between the jet power supplied and the power generated at the shaft available for coupling of the generator.

$$\eta = \frac{S.P.}{W.P.}$$

From Eqs $\eta h=\frac{R.P.}{W.P.}$, $\eta_m=\frac{S.P.}{R.P.}$ and $v_T=(P_O-P_F)VG/K$ we have

$$\eta=\eta_h\times\eta_m.$$

Classification of Turbines

The hydraulic turbines can be classified based on type of energy at the inlet, direction of flow through the vanes, head available at the inlet, discharge through the vanes and specific speed. They can be arranged as per the following table:

<table>
<tr><th colspan="2">Turbine</th><th rowspan="2">Type of energy</th><th rowspan="2">Head</th><th rowspan="2">Discharge</th><th rowspan="2">Direction of flow</th><th rowspan="2">Specific Speed</th></tr>
<tr><th>Name</th><th>Type</th></tr>
<tr><td>Pelton Wheel</td><td>Impulse</td><td>Kinetic</td><td>High Head > 250m to 1000m</td><td>Low</td><td>Tangential to runner</td><td>Low <35 Single jet
35 – 60 Multiple jet</td></tr>
<tr><td rowspan="2">Francis Turbine</td><td rowspan="3">Reaction Turbine</td><td rowspan="3">Kinetic + Pressure</td><td rowspan="2">Medium 60 m to 150 m</td><td rowspan="2">Medium</td><td>Radial flow</td><td rowspan="2">Medium 60 to 300</td></tr>
<tr><td>Mixed Flow</td></tr>
<tr><td>Kaplan Turbine</td><td>Low < 30 m</td><td>High</td><td>Axial Flow</td><td>High 300 to 1000</td></tr>
</table>

As can be seen from the above table, any specific types can be explained by suitable construction of sentences by selecting the other items in the table along the row.

Pelton wheel, named after an eminent engineer, is an impulse turbine wherein the flow is tangential to the runner and the available energy at the entrance is completely kinetic energy. Further, it is preferred at a very high head and low discharges with low specific speeds. The pressure available at the inlet and the outlet is atmospheric.

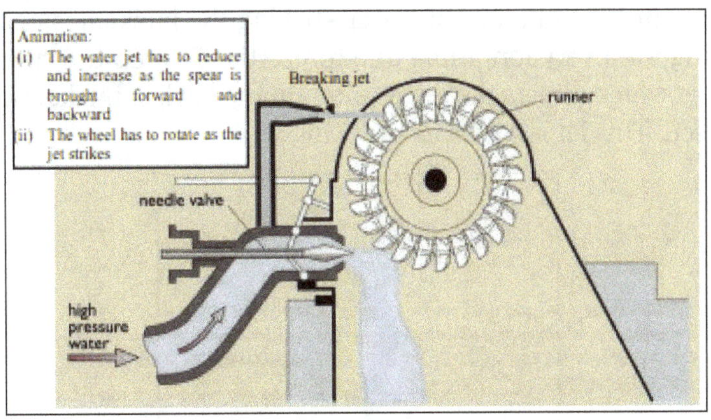

The main components of a Pelton turbine are:

(i) Nozzle and flow regulating arrangement: Water is brought to the hydroelectric plant site through large penstocks at the end of which there will be a nozzle, which converts the pressure energy completely into kinetic energy. This will convert the liquid flow into a high-speed which strikes the buckets or vanes mounted on the runner, which in-turn rotates the runner of the turbine. The amount of water striking the vanes is controlled by the forward and backward motion of the spear. As the water is flowing in the annular area between the annular area between the nozzle opening and the spear, the flow gets reduced as the spear moves forward and vice-versa.

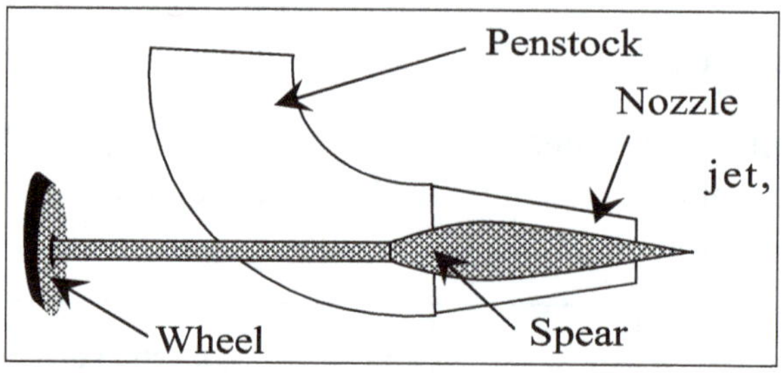

(ii) Runner with buckets: Runner is a circular disk mounted on a shaft on the periphery of which a number of buckets are fixed equally spaced as shown in Fig. The buckets are made of cast-iron cast-steel, bronze or stainless steel depending upon the head at the inlet of the turbine. The water jet strikes the bucket on the splitter of the bucket and gets deflected through (α) 160-170°.

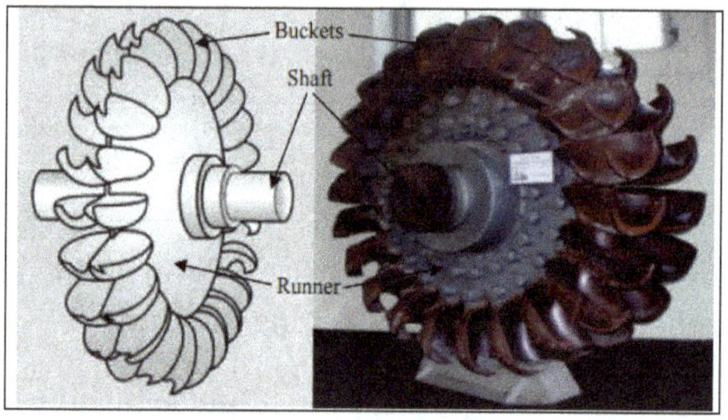

(iii) Casing: It is made of cast-iron or fabricated steel plates. The main function of the casing is to prevent splashing of water and to discharge the water into tailrace.

(iv) Breaking jet: Even after the amount of water striking the buckets is completely stopped, the runner goes on rotating for a very long time due to inertia. To stop the runner in a short time, a small nozzle is provided which directs the jet of water on the back of bucket with which the rotation of the runner is reversed. This jet is called as breaking jet.

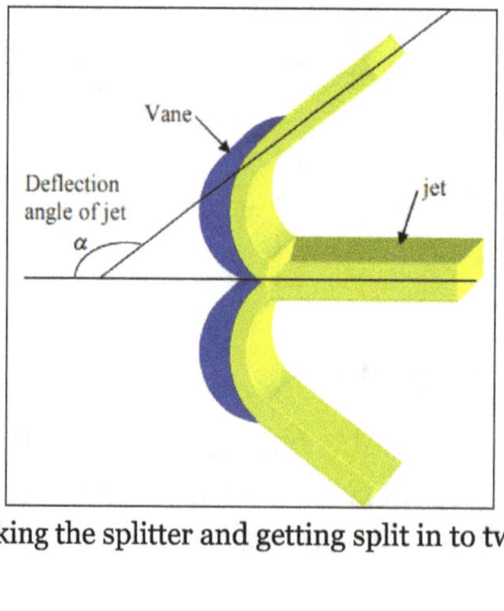

3 D Picture of a jet striking the splitter and getting split in to two parts and deviating.

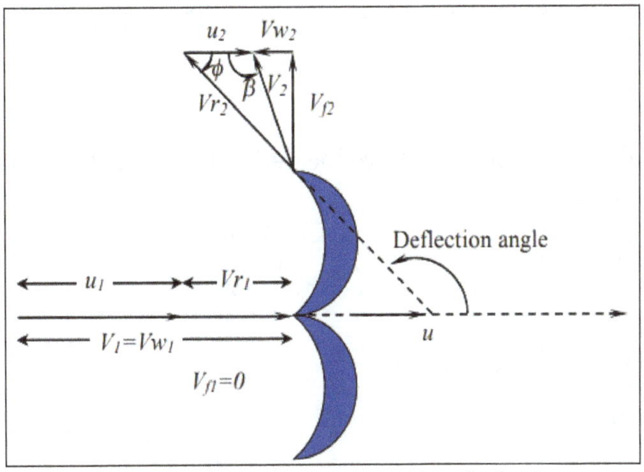

Velocity triangles for the jet striking the bucket.

From the impulse-momentum theorem, the force with which the jet strikes the bucket along the direction of vane is given by,

F_x = rate of change of momentum of the jet along the direction of vane motion,

F_x = (Mass of water/second) x change in velocity along the x direction,

$$= \rho a V_1 \left[V_{w1} - (-V_{w2}) \right]$$
$$= \rho a V_1 \left[V_{w1} + V_{w2} \right]$$

Work done per second by the jet on the vane is given by the product of Force exerted on the vane and the distance moved by the vane in one second,

$$W.D./S = Fx\ X\ u$$
$$= \rho a V_1 \left[V_{w1} + V_{w2} \right] u$$

Input to the jet per second = Kinetic energy of the jet per second,

$$= \frac{1}{2} \rho\ a V_1^3$$

$$\text{Efficiency of the jet} = \frac{\text{Output/second}}{\text{Input/second}} = \frac{\text{Workdone/second}}{\text{Input/second}}$$

$$\eta = \frac{\rho a V_1 \left[V_{w1} + V_{w2} \right] u}{\frac{1}{2} pa V_1^3}$$

$$\eta = \frac{2u \left[V_{w1} + V_{w2} \right]}{V_1^2}$$

From inlet velocity triangle, $V_{w1} = V_1$

Assuming no shock and ignoring frictional losses through the vane,

$$\text{we have } V_{r1} = V_{r2} = (V_1 - u_1)$$

In case of Pelton wheel, the inlet and outlet are located at the same radial distance from the centre of runner and hence $u_1 = u_2 = u$.

From outlet velocity triangle, we have

$$Vw_2 = V_{r2}\, Cos\phi - u_2$$
$$= (V_1 - u)\, Cos\phi - u$$

$$F_x = \rho a V_1 \left[V_1 + (V_1 - u) Cos\phi - u \right]$$
$$F_x = \rho a V_1 (V_1 - u)[1 + Cos\phi]$$

Substituting these values in the above equation for efficiency, we have

$$\eta = \frac{2u\left[V_1 + (V_1 - u)\cos\phi - u \right]}{V_1^2}$$

$$\eta = \frac{2u}{V_1^2}\left[(V_1 - u) + (V_1 - u)\cos\phi \right]$$

$$\eta = \frac{2u}{V_1^2}(V_1 - u)[1 + \cos\phi]$$

The above equation gives the efficiency of the jet striking the vane in case of Pelton wheel.

To obtain the maximum efficiency for a given jet velocity and vane angle, from maxima-minima, we have,

$$\frac{d\eta}{du} = 0$$

$$\Rightarrow \frac{d\eta}{du} = \frac{2}{V_1^2}[1 + \cos\phi]\frac{d}{du}\left(uV_1 - u^2 \right) = 0$$

$$V_1 - 2u = 0$$

$$or\ u = \frac{V_1}{2}$$

i.e. When the bucket speed is maintained at half the velocity of the jet, the efficiency of a Pelton wheel will be maximum. Substituting we get,

$$\eta_{max} = \frac{2u}{(2u)^2}(2u - u)[1 + \cos\phi]$$

$$\eta_{max} = \frac{1}{2}[1 + \cos\phi]$$

From the above it can be seen that more the value of $\cos\phi$, more will be the efficiency. Form maximum efficiency, the value of $\cos\phi$ should be 1 and the value of ϕ should be 0°. This condition makes the jet to completely deviate by 1800 and this, forces the jet striking the bucket to strike the successive bucket on the back of it acting like a breaking jet. Hence to avoid this situation, at least a small angle of ϕ = 5° should be provided.

Example:

The head at the base of the nozzle of a Pelton wheel is 640 m. The outlet vane angle of the bucket is 15°. The relative velocity at the outlet is reduced by 15 due to friction along the vanes. If the discharge at outlet is without whirl find the ratio of bucket speed to the jet speed. If the jet diameter is 100 mm while the wheel diameter is 1.2 m, find the speed of the turbine in rpm, the force exerted by the jet on the wheel, the Power developed and the hydraulic efficiency. Take C_v = 0.97.

Solution:

$H = 640$ m; $\phi = 15°$; $V_{r1} = 0.85\, V_{r2}$; $V_{w2} = 0$; d = 100 mm; D = 1.2 m;

$C_v = 0.97$; $K_u = ?$; $N = ?$; $F_x = ?$; $P = ?$; $\eta h = ?$

We know that the absolute velocity of jet is given by,

$$V = C_v\sqrt{2gH} = 0.97\,\sqrt{2\times10\times640} = 109.75\ m/s$$

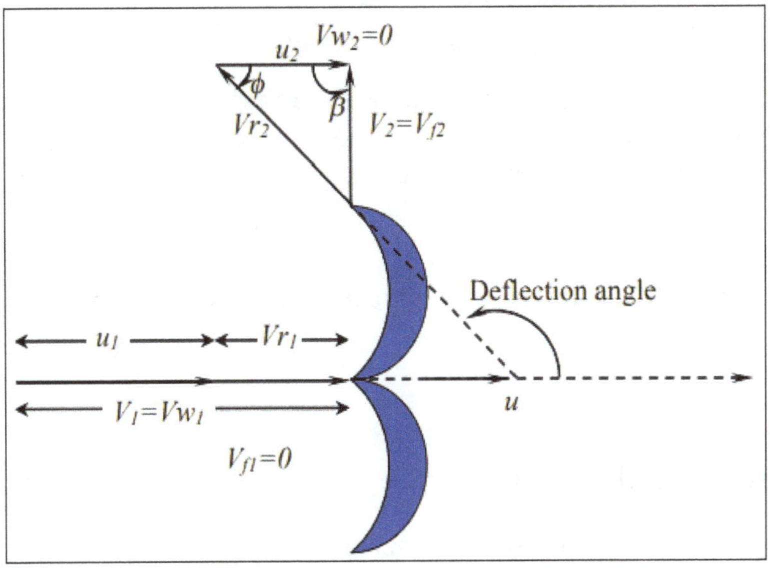

Let the bucket speed be u

Relative velocity at inlet = $V_{r1} = V_{1\text{-}u}$ = 109.74-u

Relative velocity at outlet = V_{r2} = (1-0.15) V_{r1} = 0.85(109.74 – u)

But $V_{r2}\cos\phi = u \Rightarrow 0.85(109.74 - u)\cos 15.$

Hence u = 49.48 m/s

But $u = \frac{\pi DN}{60}$ and hence,

$$N = \frac{60u}{\pi D} = \frac{60 \times 49.48}{\pi D} = 787.5 \text{ rpm}$$

Jet ratio = m = $= \frac{u}{v} = \frac{49.48}{109.74} = 0.45$

Weight of water supplied = $\gamma Q = 10 \times 1000 \times \frac{\pi}{4} \times 0.1^2 \times 109.74^2 = 8.62\ kN/s$

Force exerted = $F_x = \rho a V_1 (V_{w1} - V_{w2})$

But $V_{w1} = V_1$ and $V_{w2} = 0$ and hence,

$$F_x = 1000 \times \frac{\pi}{4} \times 0.1^2 (109.74)^2 = 94.58\ kN$$

Work done/second = F_x x u = 94.58 x 49.48 = 4679.82 kN/s

Kinetic Energy/second = $\frac{1}{2}\rho a V_1^3 = \frac{1}{2} \times 1000 \times \frac{\pi}{4} \times 0.1^2 \times 109.74^3 = 5189.85\ kN/s$

Hydraulic Efficiency = $\eta_h = \frac{Work\ done/s}{Kinetic\ Energy/s} = \frac{4679.82}{5189.85} \times 100 = 90.17\%$

Example: A PELTON wheel turbine is having a mean runner diameter of 1.0 m and is running at 1000 rpm. The net head is 100.0 m. If the side clearance is 20° and discharge is 0.1 m3 /s, find the power available at the nozzle and hydraulic efficiency of the turbine.

Solution:

D = 1.0 m; N = 1000 rpm; H = 100.0 m; ϕ = 200; Q = 0.1 m³ /s; WD/s = ?

and ηh = ?

Assume C_v = 0.98

We know that the velocity of the jet is given by,

$$V = C_v \sqrt{2gH} = 0.98\sqrt{2 \times 10 \times 1000} = 43.83\ m/s$$

The absolute velocity of the vane is given by,

$$u = \frac{\pi DN}{60} = \frac{\pi \times 1 \times 1000}{60} = 52.36\, m/s$$

This situation is impracticable and hence the data has to be modified.

Clearly state the assumption as follows:

Assume H = 700 m (Because it is assumed that the typing and seeing error as 100 for 700).

Absolute velocity of the jet is given by,

$$V = C_v\sqrt{2gH} = 0.98\sqrt{2\times10\times700} = 115.96\ m/s$$

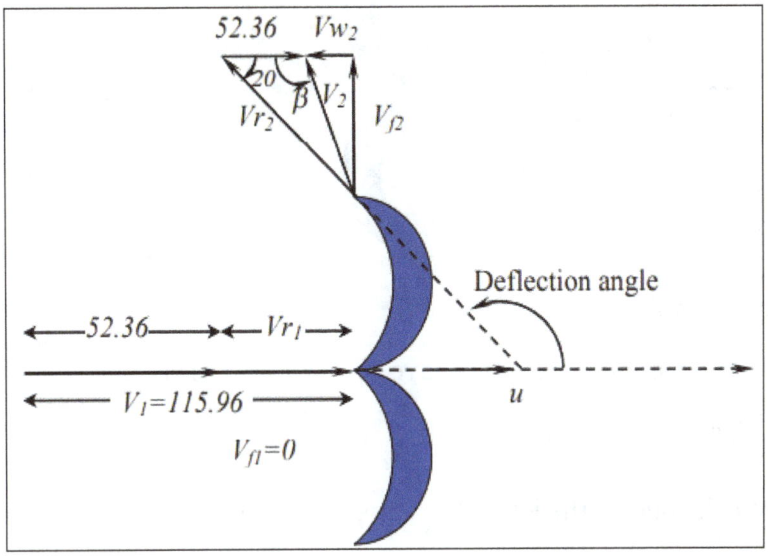

Power available at the nozzle is the given by work done per second WD/second = $\gamma\ Q H = \rho\ g\ Q\ H$ = 1000 x 10 x 0.1 x 700 = 700 kW.

Hydraulic Efficiency is given by,

$$\eta_h \frac{2u}{V_1^2}(V_1 - u)[1 + \cos\phi] = \frac{2\times52.36}{115.96^2}(115.96 - 52.36)(1 + \cos 20) = 96.07\%$$

Example:

A Pelton wheel has a mean bucket speed of 10 m/s with a jet of water flowing at the rate of 700 lps under a head of 30 m. The buckets deflect the jet through an angle of 160°. Calculate the power given by water to the runner and the hydraulic efficiency of the turbine. Assume the coefficient of nozzle as 0.98.

Solution:

$$u = 10\ m/s;\ Q = 0.7\ m^3/s;\ \phi = 180 - 160 = 20°;\ H = 30\ m;\ C_v = 0.98;$$

WD/s = ? and η_h = ?

Assume g = 10m/s²

$$V = C_v\sqrt{2gH} = 0.98\sqrt{2\times10\times30} = 24\ m/s$$

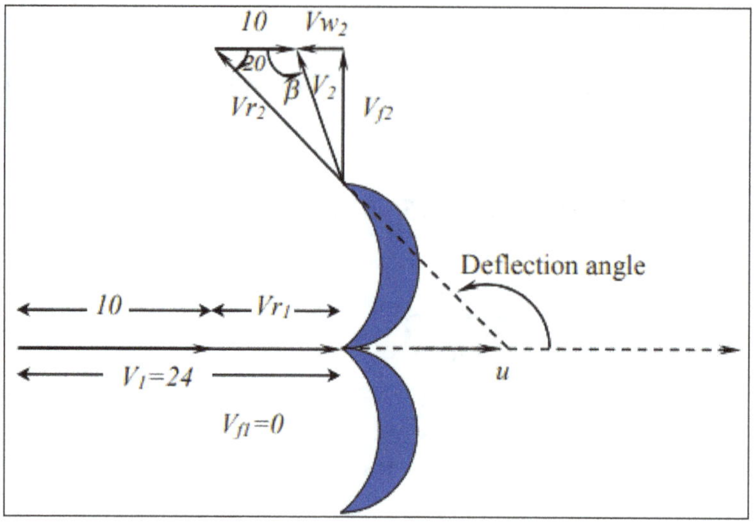

$V_{r1} = V_{1-u} = 24 - 10 = 14$ m/s

Assuming no shock and frictional losses we have $V_{r1} = V_{r2} = 14$ m/s

$$V_{w2} = V_{r2} \, Cos \, \phi - u = 14 \, x \, Cos \, 20 - 10 = 3.16 \, m/s$$

We know that the Work done by the jet on the vane is given by,

$$WD/s = \rho a V_1 \left[V_{w1} + V_{w2}\right] u = \rho Q u \left[V_{w1} + V_{w2}\right] \; as \; Q = aV_1$$

$$= 1000 \times 0.7 \times 10 \left[24 + 3.16\right] = 190.12 \; kN - m/s$$

$$IP/s = KE/s = \frac{1}{2} \rho a V_1^3 = \frac{1}{2} \rho Q V_1^2 = \frac{1}{2} \times 1000 \times 0.7 \times 24^2 = 201.6 \; kN - m/s$$

Hydraulic Efficiency = Output/ Input = 190.12/201.6 = 94.305%.

It can also be directly calculated by the derived equation as,

$$\eta_h = \frac{2u}{V_1^2}(V_1 - u)\left[1 + \cos\phi\right] = \frac{2 \times 10}{24^2}(24 - 10)\left[1 + \cos 20\right] = 94.29\%$$

A Pelton wheel has to develop 13230 kW under a net head of 800 m while running at a speed of 600 rpm. If the coefficient of Jet Cy = 0.97, speed ratio ϕ = 0.46 and the ratio of the Jet diameter is 1 /16 of wheel diameter. Calculate:

- Pitch circle diameter,
- The diameter of jet,
- The quantity of water supplied to the wheel,
- The number of Jets required.

Example: Assume overall efficiency as 85%.

Solution:

P = 13239 kW; H = 800 m; N = 600 rpm; C_v = 0.97; ϕ = 0.46 (Speed ratio)

d/D = 1/16; η_o = 0.85; D = ?; d = ?; n = ?;

Assume g = 10 m/s² and ρ = 1000 kg/m³

We know that the overall efficiency is given by,

$$\eta_o = \frac{Output}{Input} = \frac{P}{\gamma Q H} = \frac{13239 \times 10^3}{10 \times 1000 \times Q \times 800} = 0.85$$

Hence Q = 1.947 m³/s

Absolute velocity of jet is given by,

$$V = C_v \sqrt{2gH} = 0.97\sqrt{2 \times 10 \times 800} = 122.696\ m/s$$

Absolute velocity of vane is given by,

$$u = \phi\sqrt{2\,g\,H} = 0.46\sqrt{2 \times 10 \times 800} = 58.186\,m/s$$

The absolute velocity of vane is also given by,

$$u = \frac{\pi DN}{60}\ and\ hence$$

$$D = \frac{60u}{\pi N} = \frac{60 \times 58.186}{\pi \times 600} = 1.85\,m$$

$$d = \frac{1.85}{16} = 115.625\ mm$$

Discharge per jet = $q = \frac{\pi}{4} d^2 \times V = \frac{\pi}{4} \times 0.115625^2 \times 122.696 = 1.288\,m^3/s$

No. of jets = $n = \frac{Q}{q} = \frac{1.947}{1.288} \approx 2$

Example: Design a Pelton wheel for a head of 80m and speed of 300 RPM. The Pelton wheel develops 110 kW. Take co-eficient of velocity= 0.98, speed ratio= 0.48 and overall efficiency = 80%.

Solution:

H = 80 m; N = 300 rpm; P = 110 kW; C_v = 0.98, K_u=0.48; η_o = 0.80

Assume g = 10 m/s² and ρ = 1000 kg/m³

We know that the overall efficiency is given by,

$$\eta_o = \frac{Output}{Input} = \frac{P}{\gamma Q H} = \frac{110 \times 10^3}{10 \times 1000 \times Q \times 80} = 0.8$$

Hence Q = 0.171875 m³/s

Absolute velocity of jet is given by,

$$V = C_v \sqrt{2gH} = 0.98\sqrt{2 \times 10 \times 80} = 39.2\, m/s$$

Absolute velocity of vane is given by,

$$u = \phi \sqrt{2gH} = 0.45\sqrt{2 \times 10 \times 80} = 19.2\, m/s$$

The absolute velocity of vane is also given by,

$$u = \frac{\pi DN}{60} \text{ and hence}$$

$$D = \frac{60u}{\pi N} = \frac{60 \times 19.2}{\pi \times 300} = 1.22m$$

Single jet Pelton turbine is assumed.

The diameter of jet is given by the discharge continuity equation,

$$Q = \frac{\pi}{4} d^2 \times V = \frac{\pi}{4} d^2 \times 39.2 \Rightarrow 0.171875$$

Hence d = 74.7 mm.

The design parameters are Single jet Pitch Diameter = 1.22 m Jet diameter = 74.7 mm

Jet Ratio = $m = \frac{D}{d} = \frac{1.22}{0.0747} = 16.32$

No. of Buckets = 0.5×m + 15 = 24.

It is desired to generate 1000 kW of power and survey reveals that 450 m of static head and a minimum flow of 0.3m³/s are available. Comment whether the task can be accomplished by installing a Pelton wheel run at 1000 rpm and having an overall efficiency of 80%.

Hydraulic Pumps

A hydraulic pump is a mechanical device that converts mechanical power into hydraulic energy. It generates flow with enough power to overcome pressure induced by the load.

When a hydraulic pump operates, it performs two functions. First, its mechanical action creates a vacuum at the pump inlet which allows atmospheric pressure to force liquid from the reservoir into the inlet line to the pump. Second, its mechanical action delivers this liquid to the pump outlet and forces it into the hydraulic system.

A pump produces liquid movement or flow: it does not generate pressure. It produces the flow necessary for the development of pressure which is a function of resistance to fluid flow in the system. For example, the pressure of the fluid at the pump outlet is zero for a pump not connected to a system (load). Further, for a pump delivering into a system, the pressure will rise only to the level necessary to overcome the resistance of the load.

Classification of Pumps

All pumps may be classified as either positive-displacement or non-positive-displacement. Most pumps used in hydraulic systems are positive-displacement.

A non-positive-displacement pump produces a continuous flow. However, because it does not provide a positive internal seal against slippage, its output varies considerably as pressure varies. Centrifugal and propeller pumps are examples of non-positive-displacement pumps.

If the output port of a non-positive-displacement pump were blocked off, the pressure would rise, and output would decrease to zero. Although the pumping element would continue moving, flow would stop because of slippage inside the pump.

In a positive-displacement pump, slippage is negligible compared to the pump's volumetric output flow. If the output port were plugged, pressure would increase instantaneously to the point that the pump's pumping element or its case would fail (probably explode, if the drive shaft did not break first), or the pump's prime mover would stall.

Positive-displacement Principle

A positive-displacement pump is one that displaces (delivers) the same amount of liquid for each rotating cycle of the pumping element. Constant delivery during each cycle is possible because of the close-tolerance fit between the pumping element and the pump case. That is, the amount of liquid that slips past the pumping element in a positive-displacement pump is minimal and negligible compared to the theoretical maximum possible delivery. The delivery per cycle remains almost constant, regardless of changes in pressure against which the pump is working. Note that if fluid slippage is substantial, the pump is not operating properly and should be repaired or replaced.

Positive-displacement pumps can be of either fixed or variable displacement. The output of a fixed displacement pump remains constant during each pumping cycle and at a given pump speed. The output of a variable displacement pump can be changed by altering the geometry of the displacement chamber.

Other names to describe these pumps are hydrostatic for positive-displacement and hydrodynamic pumps for non-positive-displacement. Hydrostatic means that the pump converts mechanical energy to hydraulic energy with comparatively small quantity and velocity of liquid. In a hydrodynamic pump, liquid velocity and movement are large; output pressure actually depends on the velocity at which the liquid is made to flow.

Reciprocating Pumps

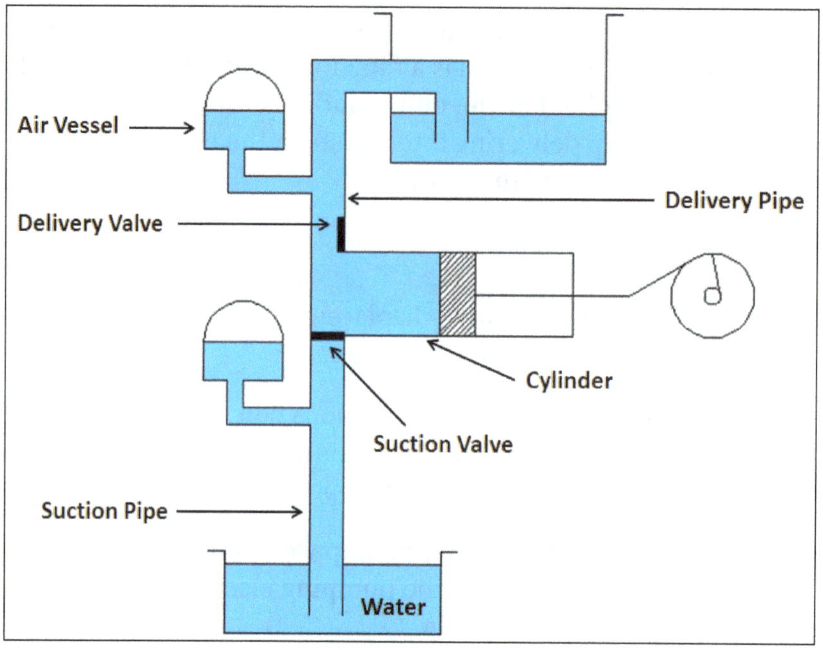

Reciprocating pump.

The positive-displacement principle is well illustrated in the reciprocating-type pump, the most elementary positive-displacement pump, Figure. As the piston extends, the partial vacuum created in the pump chamber draws liquid from the reservoir through the inlet check valve into the chamber. The partial vacuum helps seat firmly the outlet check valve. The volume of liquid drawn into the chamber is known because of the geometry of the pump case, in this example, a cylinder.

As the piston retracts, the inlet check valve reseats closing the valve, and the force of the piston unseats the outlet check valve, forcing liquid out of the pump and into the system. The same amount of liquid is forced out of the pump during each reciprocating cycle. All positive-displacement pumps deliver the same volume of liquid each cycle (regardless of whether they are reciprocating or rotating). It is a physical characteristic of the pump and does not depend on driving speed. However, the faster a pump is driven, the more total volume of liquid it will deliver.

Rotary Pumps

In a rotary-type pump, rotary motion carries the liquid from the pump inlet to the pump outlet. Rotary pumps are usually classified according to the type of element that transmits the liquid, so that we speak of a gear-, lobe-, vane-, or piston-type rotary pump.

External-gear pumps can be divided into external and internal-gear types. A typical external-gear pump is shown in figure. These pumps come with a straight spur, helical, or herringbone gears.

Straight spur gears are easiest to cut and are the most widely used. Helical and herringbone gears run more quietly, but cost more.

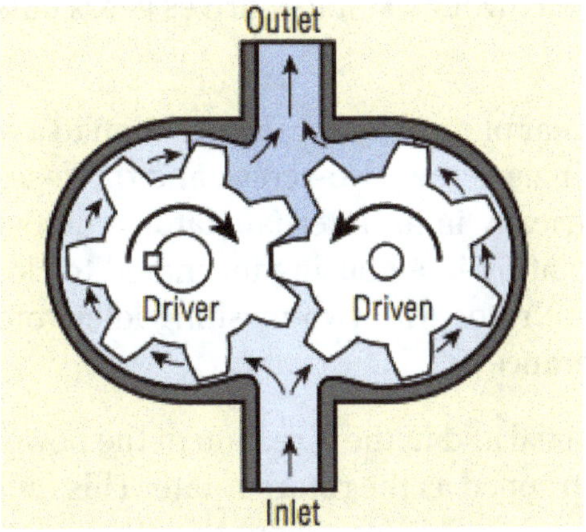

Spur gear pump

A gear pump produces flow by carrying fluid in between the teeth of two meshing gears. One gear is driven by the drive shaft and turns the idler gear. The chambers formed between adjacent gear teeth are enclosed by the pump housing and side plates (also called wear or pressure plates).

A partial vacuum is created at the pump inlet as the gear teeth unmesh. Fluid flows in to fill the space and is carried around the outside of the gears. As the teeth mesh again at the outlet end, the fluid is forced out.

Volumetric efficiencies of gear pumps run as high as 93% under optimum conditions. Running clearances between gear faces, gear tooth crests and the housing create an almost constant loss in any pumped volume at a fixed pressure. This means that volumetric efficiency at low speeds and flows is poor, so that gear pumps should be run close to their maximum rated speeds.

Although the loss through the running clearances, or "slip," increases with pressure, this loss is nearly constant as speed and output change. For one pump the loss increases by about 1.5 gpm from zero to 2,000 psi regardless of speed. Change in slip with pressure change has little effect on performance when operated at higher speeds and outputs. External-gear pumps are comparatively immune to contaminants in the oil, which will increase wear rates and lower efficiency, but sudden seizure and failure are not likely to occur.

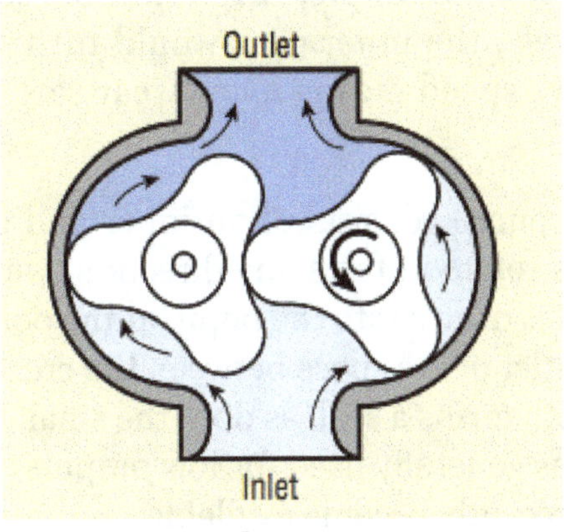

Lobe pump

The lobe pump is a rotary, external-gear pump, Figure. It differs from the conventional external-gear pump in the way the "gears" are driven. In a gear pump, one gear drive the other; in a lobe pump, both lobes are driven through suitable drives gears outside of the pump casing chamber.

A screw pump is an axial-flow gear pump, similar in operation to a rotary screw compressor. Three types of screw pumps are the single-screw, two-screw, and three-screw. In the single-screw pump, a spiraled rotor rotates eccentrically in an internal stator. The two-screw pump consists of two parallel intermeshing rotors rotating in a housing machined to close tolerances. The three-screw pump consists of a central-drive rotor with two meshing idler rotors; the rotors turn inside of a housing machined to close tolerances.

Flow through a screw pump is axial and in the direction of the power rotor. The inlet hydraulic fluid that surrounds the rotors is trapped as the rotors rotate. This fluid is pushed uniformly with the rotation of the rotors along the axis and is forced out the other end. The fluid delivered by a screw pump does not rotate, but moves linearly. The rotors work like endless pistons, which continuously move forward. There are no pulsations even at higher speed. The absence of pulsations and the fact that there is no metal-to-metal contact results in very quiet operation. Larger pumps are used as low-pressure, large-volume prefill pumps on large presses. Other applications include hydraulic systems on submarines and other uses where noise must be controlled.

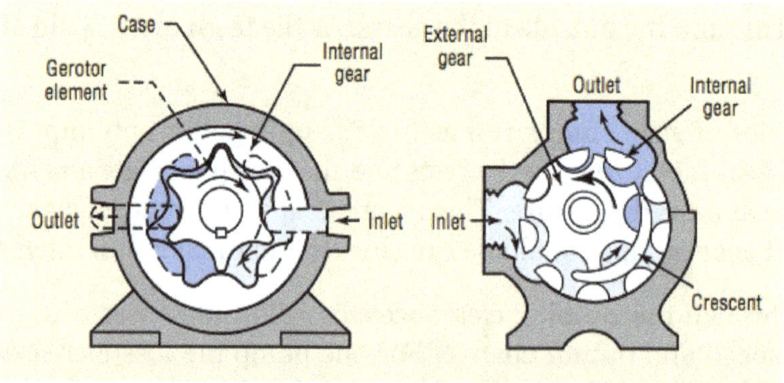

Internal-gear pumps - gerotor and crescent

Internal-gear pumps, have an internal gear and an external gear. Because these pumps have one or two less teeth in the inner gear than the outer, relative speeds of the inner and outer gears in these designs are low. For example, if the number of teeth in the inner and outer gears were 10 and 11 respectively, the inner gear would turn 11 revolutions, while the outer would turn 10. This low relative speed means a low wear rate. These pumps are small, compact units.

The crescent seal internal-gear pump consists of an inner and outer gear separated by a crescent-shaped seal. The two gears rotate in the same direction, with the inner gear rotating faster than the outer. The hydraulic oil is drawn into the pump at the point where the gear teeth begin to separate and is carried to the outlet in the space between the crescent and the teeth of both tears. The contact point of the gear teeth forms a seal, as does the small tip clearance at the crescent. Although in the past this pump was generally used for low outputs, with pressures below 1,000 psi, a 2-stage, 4,000-psi model has recently become available.

The gerotor internal-gear pump consists of a pair of gears which are always in sliding contact. The internal gear has one more tooth than the gerotor gear. Both gears rotate in the same direction. Oil is drawn into the chamber where the teeth are separating, and is ejected when the teeth start to mesh again. The seal is provided by the sliding contact.

Generally, the internal-gear pump with toothcrest pressure sealing has higher volumetric efficiency at low speeds than the crescent type. Volumetric and overall efficiencies of these pumps are in the same general range as those of external-gear pumps. However, their sensitivity to dirt is somewhat higher.

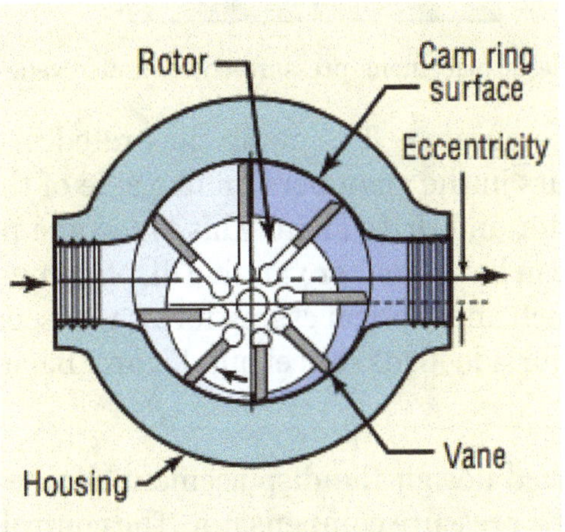

Basic (unbalanced) vane pump.

In vane pumps, a number of vanes slide in slots in a rotor which rotates in a housing or ring. The housing may be eccentric with the center of the rotor, or its shape may be oval, Figure. In some designs, centrifugal force holds the vanes in contact with the housing, while the vanes are forced in and out of the slots by the eccentricity of the housing. In one vane pump, light springs hold the vanes against the housing; in another pump design, pressurized pins urge the vanes outward.

During rotation, as the space or chamber enclosed by vanes, rotor, and housing increases, a vacuum is created, and atmospheric pressure forces oil into this space, which is the inlet side of the pump. As the space or volume enclosed reduces, the liquid is forced out through the discharge ports.

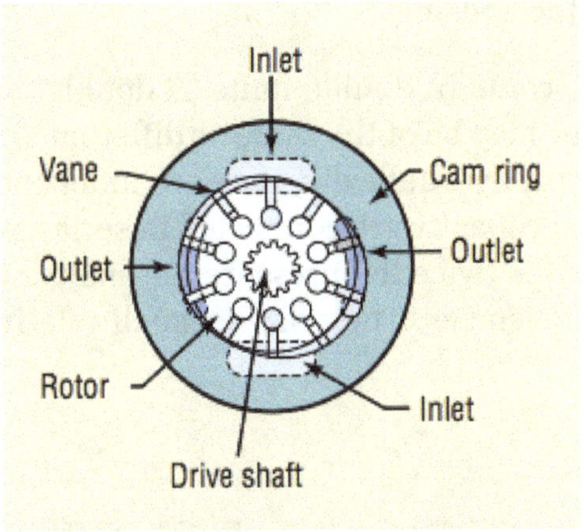

Balanced vane pump.

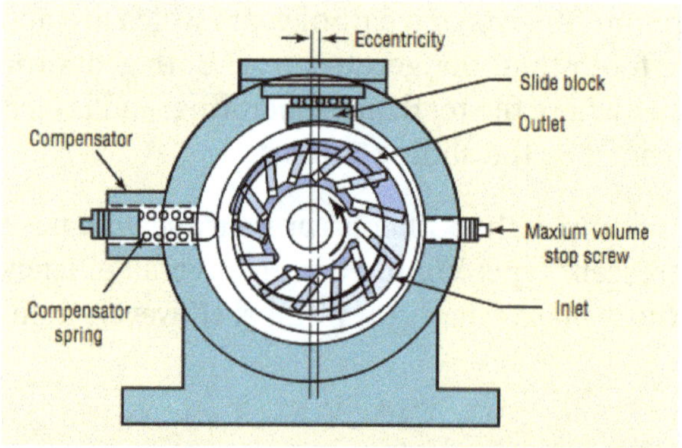

Variable-displacement, pressure-compensated vane pump.

Balanced and unbalanced vane pumps— The pump illustrated in figure is unbalanced, because all of the pumping action occurs in the chambers on one side of the rotor and shaft. This design imposes a side load on the rotor and drive shaft. This type vane pump has a circular inner casing. Unbalanced vane pumps can have fixed or variable displacements. Some vane pumps provide a *balanced* construction in which an elliptical casing forms two separate pumping areas on opposite sides of the rotor, so that the side loads cancel out, Figure. Balanced vane pumps come only in fixed displacement designs.

In a variable-volume unbalanced design the displacement can be changed through an external control such as a handwheel or a pressure compensator. The control moves the cam ring to change the eccentricity between the ring and rotor, thereby changing the size of the pumping chamber and thus varying the displacement per revolution.

When pressure is high enough to overcome the compensator spring force, the cam ring shifts to decrease the eccentricity. Adjustment of the compensator spring determines the pressure at which the ring shifts. Because centrifugal force is required to hold the vanes against the housing and maintain a tight seal at those points, these pumps are not suited for low-speed service. Operation at speeds below 600 rpm is not recommended. If springs or other means are used to hold vanes out against the ring, efficient operation at speeds of 100 to 200 rpm is possible.

Vane pumps maintain their high efficiency for a long time, because compensation for wear of the vane ends and the housing is automatic. As these surfaces wear, the vanes move further out in their slots to maintain contact with the housing.

Vane pumps, like other types, come in double units. A double pump consists of two pumping units in the same housing. They may be of the same or different sizes. Although they are mounted and driven like single pumps, hydraulically, they are independent. Another variation is the series unit: two pumps of equal capacity are connected in series, so that the output of one feeds the other. This arrangement gives twice the pressure normally available from this pump. Vane pumps have relatively high efficiencies. Their size is small relative to output. Dirt tolerance is relatively good.

Piston Pumps

The piston pump is a rotary unit which uses the principle of the reciprocating pump to produce fluid flow. Instead of using a single piston, these pumps have many piston-cylinder combinations. Part of the pump mechanism rotates about a drive shaft to generate the reciprocating motions, which draw fluid into each cylinder and then expels it, producing flow. There are two basic types, axial and radial piston; both area available as fixed and variable displacement pumps. The second variety often is capable of variable reversible (overcenter) displacement.

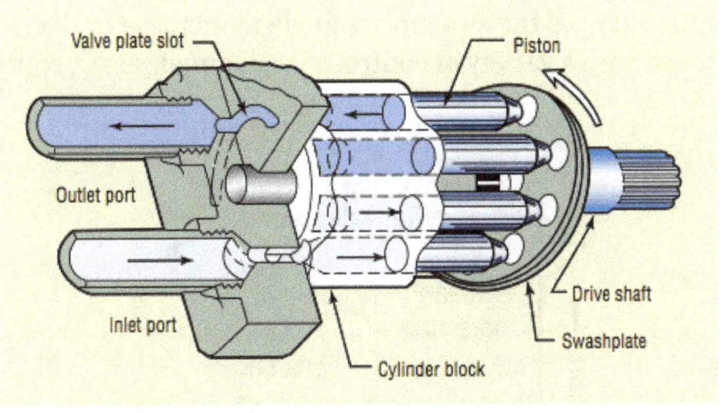

Axial-piston pump varies displacement by changing angle of swashplate.

Most axial and radial piston pumps lend themselves to variable as well as fixed displacement designs. Variable displacement pumps tend to be somewhat larger and heavier, because they have added internal controls, such as handwheel, electric motor, hydraulic cylinder, servo, and mechanical stem.

Axial-piston pumps: The pistons in an axial piston pump reciprocate parallel to the centerline of the drive shaft of the piston block. That is, rotary shaft motion is converted into axial reciprocating motion. Most axial piston pumps are multi-piston and use check valves or port plates to direct liquid flow from inlet to discharge.

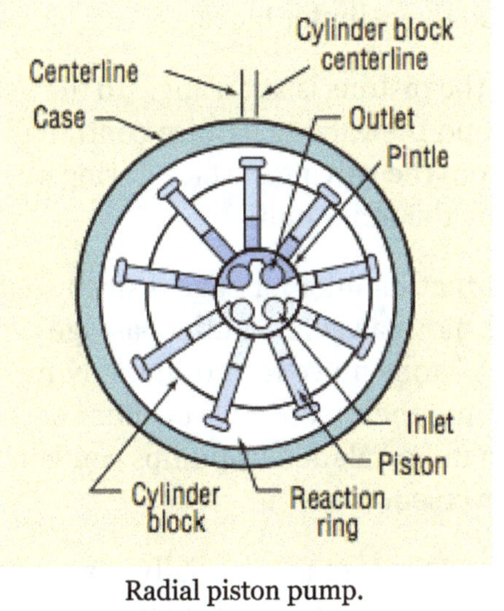

Radial piston pump.

Inline piston pumps: The simplest type of axial piston pump is the swashplate design in which a cylinder block is turned by the drive shaft. Pistons fitted to bores in the cylinder block are connected through piston shoes and a retracting ring, so that the shoes bear against an angled swashplate.

As the block turns, figure, the piston shoes follow the swashplate, causing the pistons to reciprocate. The ports are arranged in the valve plate so that the pistons pass the inlet as they are pulled out and the outlet as they are forced back in. In these pumps, displacement is determined by the size and number of pistons as well as their stroke length, which varies with the swashplate angle.

In variable-displacement models of the inline pump, the swashplate swings in a movable yoke. Pivoting the yoke on a pintle changes the swashplate angle to increase or decrease the piston stroke. The yoke can be positioned with a variety of controls, *i.e.*, manual, servo, compensator, handwheel, etc.

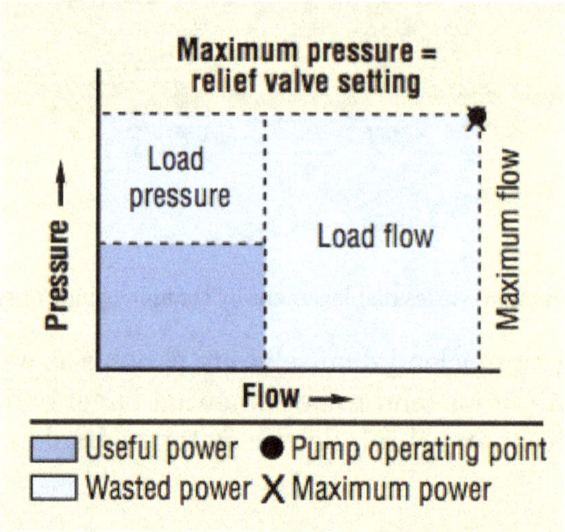

Pressure-flow curve of fixed-displacement hydraulic pump.

Bent-axis pumps: This pump consists of a drive shaft which rotates the pistons, a cylinder block, and a stationary valving surface facing the cylinder block bores which ports the inlet and outlet flow. The drive shaft axis is angular in relation to the cylinder block axis. Rotation of the drive shaft causes rotation of the pistons and the cylinder block.

Because the plane of rotation of the pistons is at an angle to the valving surface plane, the distance between any one of the pistons and the valving surface continually changes during rotation. Each individual piston moves away from the valving surface during one-half of the shaft revolution and toward the valving surface during the other half.

The valving surface is so ported that its inlet passage is open to the cylinder bores in that part of the revolution where the pistons move away. Its outlet passage is open to the cylinder bores in the part of the revolution where the pistons move toward the valving surface. Therefore, during pump rotation the pistons draw liquid into their respective cylinder bores through the inlet chamber and force it out through the outlet chamber. Bent axis pumps come in fixed and variable displacement configurations, but cannot be reversed.

In radial-piston pumps, the pistons are arranged radially in a cylinder block; they move perpendicularly to the shaft centerline. Two basic types are available: one uses cylindrically shaped pistons, the other ball pistons. They may also be classified according to the porting arrangement: check valve or pintle valve. They are available in fixed and variable displacement, and variable reversible (over-center) displacement.

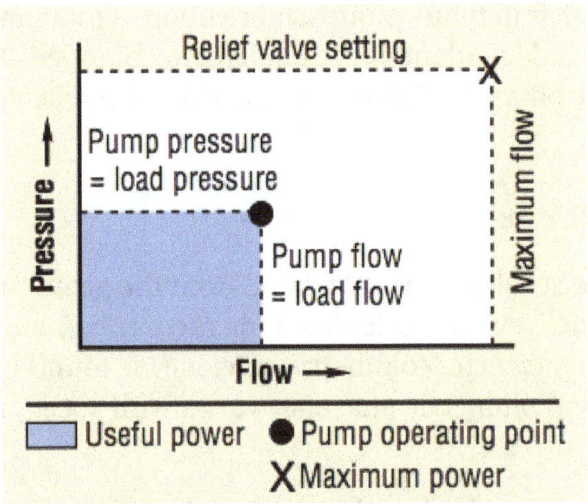

Pressure flow curve of variable-displacement hydraulic pump with ideal flow and pressure compensation.

In pintle-ported radial piston pump, figure, the cylinder block rotates on a stationary pintle and inside a circular reacting ring or rotor. As the block rotates, centrifugal force, charging pressure, or some form of mechanical action causes the pistons to follow the inner surface of the ring, which is offset from the centerline of the cylinder block. As the pistons reciprocate in their bores, porting in the pintle permits them to take in fluid as they move outward and discharge it as they move in.

The size and number of pistons and the length of their stroke determine pump displacement. Displacement can be varied by moving the reaction ring to increase or decrease piston travel, varying eccentricity. Several controls are available for this purpose.

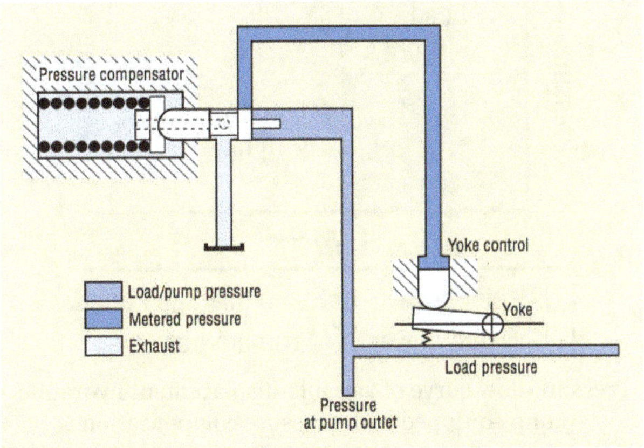

Schematic of typical proportional pump pressure compensator control.

Plunger pumps are somewhat similar to rotary piston types, in that pumping is the result of pistons reciprocating in cylinder bores. However, the cylinders are fixed in these pumps; they do not rotate around the drive shaft. Pistons may be reciprocated by a crankshaft, by eccentrics on a shaft, or by a wobble plate. When eccentrics are used, return stroke is by springs. Because valving cannot be supplied by covering and uncovering ports as rotation occurs, inlet and outlet check valves may be used in these pumps.

Because of their construction, these pumps offer two features other pumps do not have: one has a more positive sealing between inlet and outlet, permitting higher pressures without excessive leakage of slip. The other is that in many pumps, lubrication of moving parts other than the piston and cylindrical bore may be independent of the liquid being pumped. Therefore, liquids with poor lubricating properties can be pumped. Volumetric and overall efficiencies are close to those of axial and radial piston pumps.

Measuring Pump Performance

Volume of fluid pumped per revolution is calculated from the geometry of the oil-carrying chambers. A pump never quite delivers the calculated, or *theoretical,* amount of fluid. How close it comes is called *volumetric efficiency*. Volumetric efficiency is found by comparing the calculated delivery with actual delivery. Volumetric efficiency varies with speed, pressure, and the construction of the pump.

A pump's *mechanical efficiency* is also less than perfect, because some of the input energy is wasted in friction. Overall efficiency of a hydraulic pump is the product of its volumetric and mechanical efficiencies.

Pumps are generally rated by their maximum operating pressure capability and their output, in gpm or lpm, at a given drive speed, in rpm.

Matching Pump Power with the Load

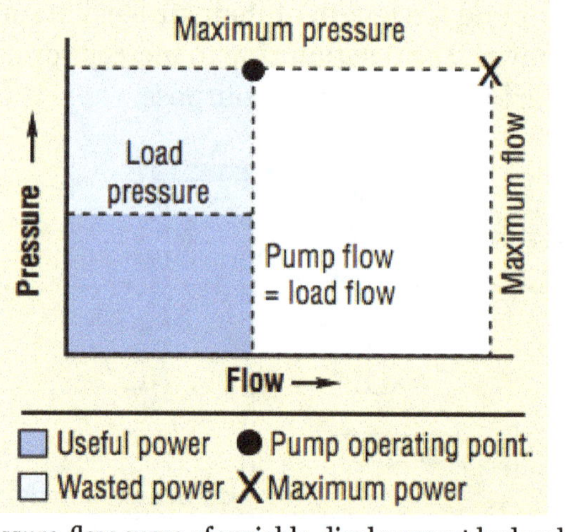

Pressure-flow curve of variable-displacement hydraulic pump equipped with pressure compensation.

Pressure compensation and load sensing are terms often used to describe pump features that improve the efficiency of pump operation. Sometimes these terms are used interchangeably, a misconception that is cleared up once you understand the differences in how the two enhancements operate. To investigate these differences, consider a simple circuit using a fixed-displacement pump running at constant speed. This circuit is efficient only when the load demands maximum power because the pump puts out full pressure and flow regardless of load demand. A relief valve prevents excessive pressure buildup by routing high-pressure fluid to tank when the system reaches the relief setting. As figure shows, power is wasted whenever the load requires less than full flow or full pressure. The unused fluid energy produced by the pump becomes heat that must be dissipated. Overall system efficiency may be 25% or lower.

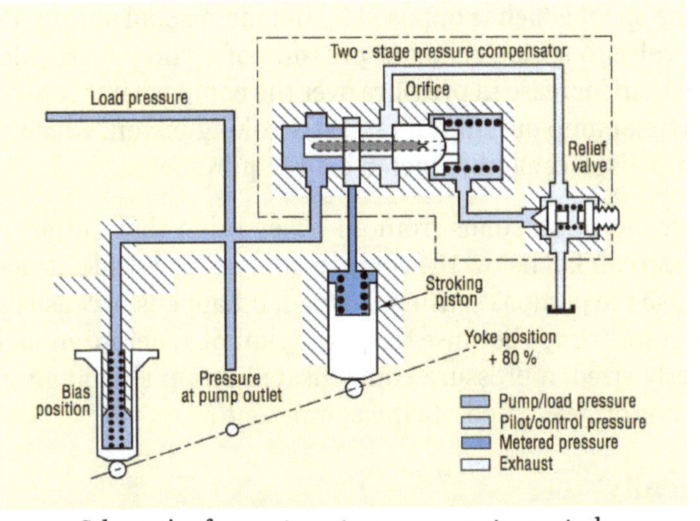

Schematic of pump two-stage compensator control.

Variable displacement pumps, equipped with displacement controls, Figure, can save most of this wasted hydraulic horsepower when moving a single load. Control variations include hand wheel, lever, cylinder, stem servo, and electrohydraulic servo controls. Examples of displacement control applications are the lever-controlled hydrostatic transmissions used to propel windrowers, skid-steer loaders, and road rollers.

While matching the exact flow and pressure needs of a single load, these controls have no inherent pressure or power-limiting capabilities. And so, other provisions must be made to limit maximum system pressure, and the prime mover still must have corner horsepower capability. Moreover, when a pump supplies a circuit with multiple loads, the flow and pressure-matching characteristics are compromised.

A design approach to the system in which one pump powers multiple loads is to use a pump equipped with a proportional pressure compensator, Figure. A yoke spring biases the pump swash plate toward full displacement. When load pressure exceeds the compensator setting, pressure force acts on the compensator spool to overcome the force exerted by the spring.

The spool then shifts toward the compensator-spring chamber, ports pump output fluid to the stroking piston, and decreases pump displacement. The compensator spool returns to neutral when pump pressure matches the compensator spring setting. If a load blocks the actuators, pump flow drops to zero.

Using a variable-displacement, pressure-compensated pump rather than a fixed-displacement pump reduces circuit horsepower requirements dramatically, Figure. Output flow of this type of pump varies according to a predetermined discharge pressure as sensed by an orifice in the pump's compensator. Because the compensator itself operates from pressurized fluid, the discharge pressure must be set higher - say, 200 psi higher - than the maximum load-pressure setting. So if the load-pressure setting of a pressure-compensated pump is 1,100 psi, the pump will increase or decrease its displacement (and output flow) based on a 1,300-psi discharge pressure.

A two-stage pressure-compensator control, Figure, uses pilot flow at load pressure across an orifice in the main stage compensator spool to create a pressure drop of 300 psi. This pressure drop generates a force on the spool which is opposed by the main spool spring. Pilot fluid flows to tank through a small relief valve. A spring chamber pressure of 4,700 psi provides a compensator control setting of 5,000 psi. An increase in pressure over the compensator setting shifts the main stage spool to the right, porting pump output fluid to the stroking piston, which overcomes bias piston force and reduces pump displacement to match load requirements.

The earlier stated misconception stems from an observation that output pressure from a pressure-compensated pump can fall below the compensator setting while an actuator is moving. This does not happen because the pump is sensing the load, it happens because the pump is undersized for the application. Pressure drops because the pump cannot generate enough flow to keep up with the load. When properly sized, a pressure-compensated pump should always force enough fluid through the compensator orifice to operate the compensator.

Superior Dynamically

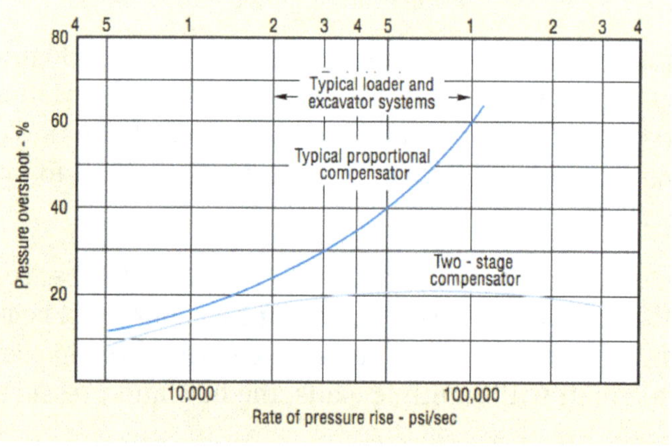

Typical performance of a single- and two-stage pressure compensation.

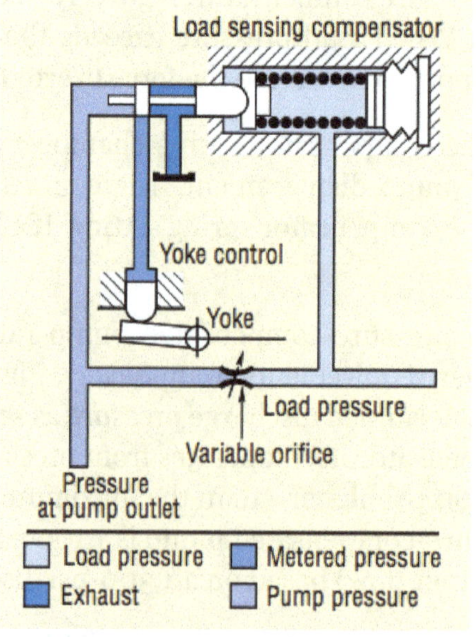

Schematic of proportional pump compensator that provides load-sensing capability.

With respect to its matching function, a two-stage compensator is identical to the proportional compensator control shown in Figure. The dynamic performance of the two-stage control is superior, however. This becomes obvious when one analyzes a transient which involves a sudden decrease in load flow demand, starting from full stroke at low pressure.

The single-stage control spool ports pressure fluid to the stroke piston only when pump discharge pressure reach the compensator setting. The main-stage spool of the two-stage control starts moving as soon as pump discharge pressure minus spring chamber pressure exceeds the 300-psi spring setting. Because pilot fluid flows through the orifice and because of the flow needed to compress the fluid in the spring chamber, the spring chamber pressure lags pump discharge pressure. This causes the spool to become unbalanced and shift to the right.

Pump destroking starts before pump discharge pressure reaches the compensator setting, Figure. Note that in system equipped with an accumulator, a two-stage compensator control provides little advantage. In excavator hydraulic systems, however, superiority of the two-stage compensator is evident: it provides system components much greater protection against pressure transients.

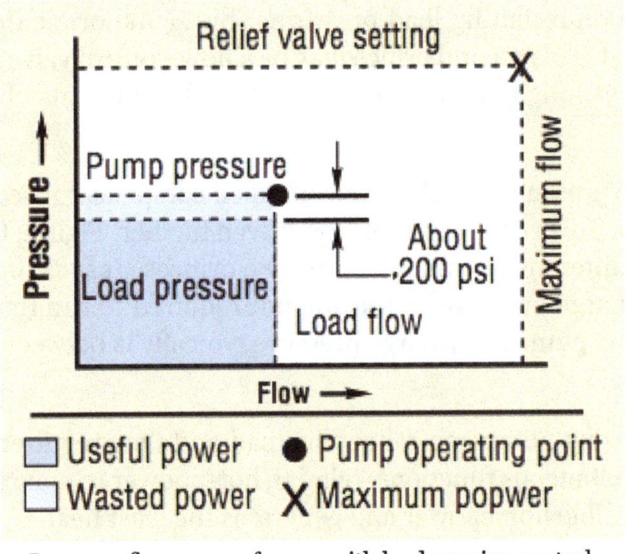

Pressure-flow curve of pump with load-sensing control.

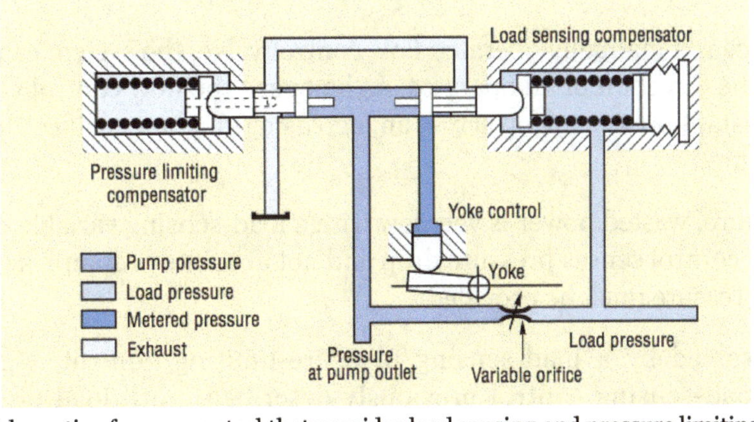

Schematic of pump control that provides load sensing and pressure limiting.

Load Sensing: The Next Step

A similar control, which has recently become popular, is the *load sensing* control, sometimes called a power matching control, Figure. The single-stage valve is almost identical to the single-stage compensator control, Figure, except that the spring chamber is connected downstream of a variable orifice rather than directly to tank. The load-sensing compensator spool achieves equilibrium when the pressure drop across the variable orifice matches the 300-psi spring setting.

Any of three basic load-sensing signals control a load-sensing pump: unloaded, working, and relieving. In the unloaded mode, the lack of load pressure causes the pump to produce zero discharge flow at bias or unload pressure. When working load pressure causes the pump to generate discharge flow in relation to a set pressure drop or bias pressure. When the system reaches maximum pressure, the pump maintains this pressure by adjusting its discharge flow.

Like the pressure-compensated pump, a load-sensing pump has a pressure-compensation control, but the control is modified to receive two pressure signals, not just one. As with pressure compensation, the load-sensing control receives a signal representing discharge pressure, but it also receives a second signal representing load pressure. This signal originates from a second orifice downstream from the first. This second orifice may be a flow-control valve immediately beyond the pump outlet, the spool opening of a directional control valve, or it may be a restriction in a fluid conductor.

Comparison of these two pressure signals in the modified compensator section allows the pump to sense both load and flow. This reduces power losses even further, Figure. Output flow of the pump varies in relation to the differential pressure of the two orifices. Just as the pressure-compensated pump increased its discharge pressure by the amount required to run the pressure compensator, the load- and flow-sensing pump's discharge pressure typically is between 200 and 250 psi higher than actual load pressure.

Furthermore, a load-sensing pump can follow the load and flow requirements of a single circuit function or multiple simultaneous functions, relating horsepower to maximum load pressure. This consumes the lowest possible horsepower and generates the least heat.

Operator Control

If the variable orifice is a manually operated flow control valve, the system can operate in a load-matched mode at the direction of an operator. As he opens the flow control valve, flow increases proportionally (constant pressure drop across an increasing-diameter orifice), at a pressure slightly above load pressure.

As suggested in Figure, wasted power is very low with a load-sensing variable volume pump compensator. Since the control senses pressure drop and not absolute pressure, a relief valve or other means of limiting pressure must be provided.

This problem is solved by a load-sensing/pressure-limiting control, Figure. This control functions as the load-sensing control previously described, until load pressure reaches the pressure limiter setting. At that point, the limiter portion of the compensator overrides the load-sensing control to destroke the pump. Again, the prime mover must have corner horsepower capability.

Load-Sensing Gear Pumps

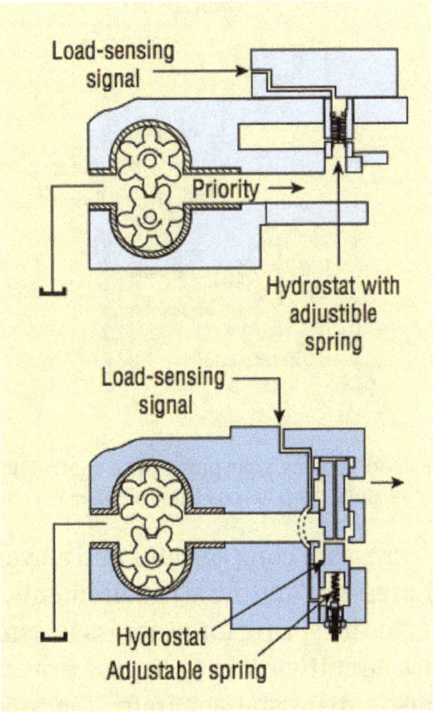

Load-sensing gear pumps with two different types of hydrostats installed. The spring adjustment allows tuning pressure drop for different manufacturers' valves or line lengths.

Piston and vane pumps rely on their variable-displacement capability to accomplish load sensing. How, then, can a gear pump accomplish load sensing if its displacement is fixed? Like standard gear pumps, load-sensing gear pumps have low initial cost when compared to other designs with equivalent flow and pressure capabilities. However, load-sensing gear pumps offer the versatility of variable-displacement axial-piston and vane pumps but without the high complexity and high cost of variable-displacement mechanisms.

A load-sensing gear pump can:

- Provide the high efficiency of load sensing without the high cost associated with piston or vane pumps,
- Produce zero to full output flow in less than 40 milliseconds with little or no pressure spiking and without pump inlet supercharging,
- Drive circuits with low (approaching atmospheric) unload relief pressures,
- Provide priority flow and secondary flow with a low unload pressure to reduce standby and secondary loaded power draw, and
- Interchange with load-sensing vane or piston pumps without having to change line or component sizes.

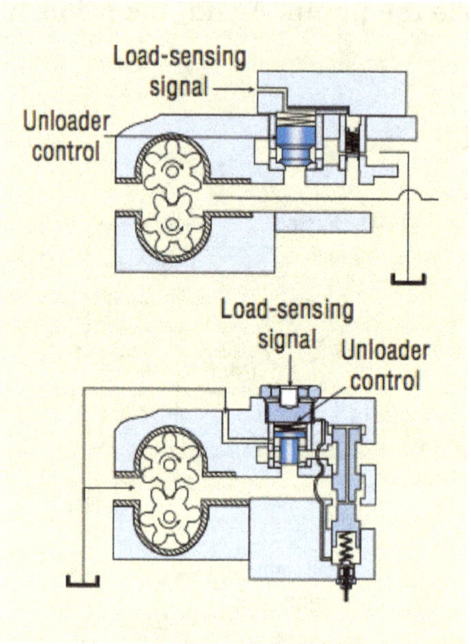

Unloader control has been added to the load-sensing gear pump. The control uses a poppet or a plunger to allow maximum flow at the minimum pressure drop across the unloader with minimal control movement.

Load-sensing piston pumps use a pressure compensator and a hydrostat to vary volumetric output to a system in reference to load pressure and flow requirements. A hydrostat is a spring-loaded device that meters flow according to the spring force across its equal but opposing effective areas. It may be restrictive, as in a series circuit, or it may bypass primary load pressure to a secondary or tank pressure. In simple terms, a hydrostat separates the total flow into two flows: one represents the required flow and the other represents the required pressure of the primary circuit. A load-sensing piston pump uses its hydrostat to regulate output flow relative to load pressure and bypasses the excess pump flow to a secondary route, which may be ported to tank or to a secondary circuit.

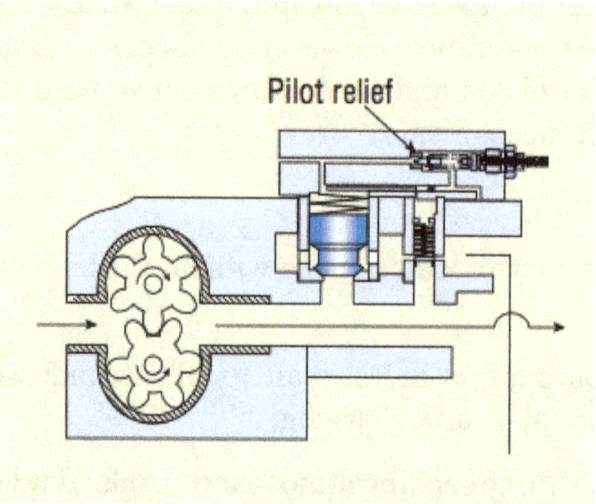

Combined control is achieved by incorporating a pilot relief, which causes the hydrostat to act as the main stage of a pilot-operated relief valve.

A load-sensing gear pump, on the other hand, uses a hydrostat in combination with an unloader to vary its volumetric output in response to load and flow requirements. Because load-sensing piston and gear pumps both use a single load-sensing signal to control pump discharge pressure and flow, they are interchangeable in load-sensing circuits. Both types have much in common and offer substantial power savings over systems using fixed-displacement pumps. Both offer reduced power consumption in the running mode, when flow and pressure are required to operate a function. They also conserve power in the standby mode, when the system is idling or in a non-operational mode. Furthermore, they can reduce the required size and therefore, cost of valves, conductors, and filters needed for the circuit.

The load-sensing gear pump illustrated in figure minimizes power consumption in the running mode by separating total discharge flow according to a remote primary function pressure and a primary flow. This is accomplished through a single load-sensing signal originating from the priority circuit and routed as close as possible to the discharge side of the pump's gears.

Adding an unloader control to the pump circuit, figure, allows the system to conserve power in the standby mode of operation as well as in the running mode. This control must be installed in parallel with the inlet port of the hydrostat and as close as possible to the discharge side of the gears. It must be piloted by the same load-sensing signal as in Figure. This signal causes the pump to dump all flow from the outlet to the secondary circuit and at a pressure well below the hydrostat's pressure-drop setting in the standby mode.

The unloader control must operate off the same remote load-sensing signal that controls the hydrostat. Unlike the hydrostat, the unloader poppet of the unloader control is designed with opposing areas having a ratio of at least 2:1. Any line pressure sensed that exceeds 50% of pump discharge pressure will close the unloader control. The ability of the unloader control to unload the pump to near atmospheric discharge pressure is controlled by the poppet or plunger spring force. The unloader control is set to the lowest value to maintain the internal pressure loading of the gear pump. When compared to a standard fixed-displacement gear pump circuit, this control can reduce standby power consumption by 90%.

Dual and Combined Controls

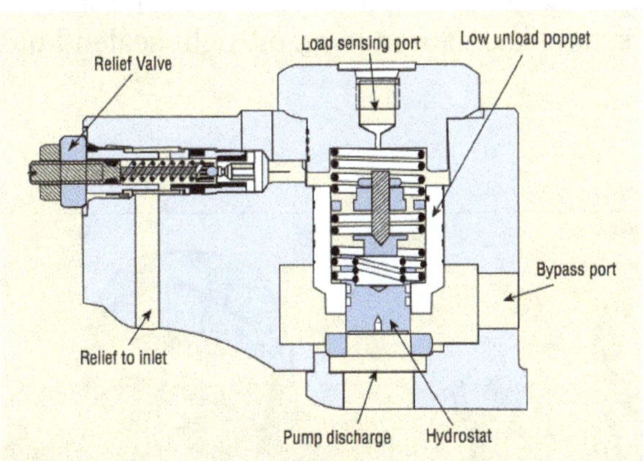

This cutaway shows combined control, which has an adjustable hydrostat contained within the unloader control. Locating the hydrostat within the low-unload control allows all piston areas to operate from a single load-response signal. It is intended for applications using large pumps where secondary flow bypasses to tank.

The load-sensing signal can be conditioned by limiting pressure in the remote sensing line or taking it to 0 psig. Doing so causes the hydrostat and the unloader control of the load-sensing gear pump to respond to the conditioned signal according to the discharge pressure. This is accomplished by providing a pilot relief, figure, which causes the hydrostat to act as the main stage of a pilot-operated relief valve. The ability to condition the load-sensing line is patented and makes the load-sensing gear pump useful for functions other than just load sensing.

The combined-control load-sensing gear pump, figure, is intended for large-displacement pumps and bypasses secondary flow to tank. It also is patented, and can be used in the same applications as the dual-control pump. However, because secondary flow must be routed to tank, it cannot be used when the secondary circuit drives a load.

Hydraulic Coupling

Fluid coupling is also known as hydraulic coupling is a hydrodynamic device which is used to transfer rotational power from one shaft to another by the use of transmission fluid. It is used in automotive transmission system, marine propulsion system and in industries for power transmission. It is used as an alternative for the mechanical clutch.

It consists of three main components:

1. Housing: It is also known as the shell. It has oil-tight seal around the drive shaft. It also protects the impeller and turbine from outside damage.
2. Impeller or pump: It is a turbine which is connected to the input shaft and called as impeller. It is also known as pump because it acts as a centrifugal pump.
3. Turbine: It is connected to the output shaft to which the rotational power is to be transmitted.

The impeller is connected to the prime mover (internal combustion engine) which is a power source. The turbine is connected to the output shaft where rotation power is needed to be transmitted. The impeller and turbine is enclosed in an oil-tight sealed housing. The housing consists of transmission fluid.

Working Principle

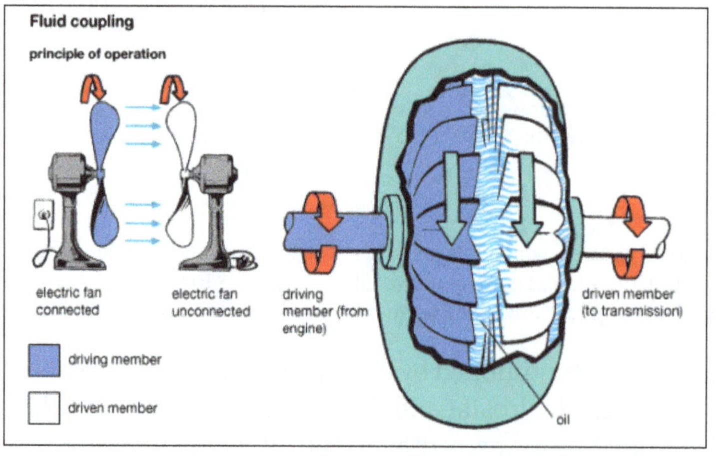

The working principle of fluid can be easily explained by the taking two fans in which one is connected to the power supply and other is not. When the power switch is ON, the air from the first fan is starts to blow towards the second fan. Initially when the first fan is blowing at lower speed, it does not able to drive the second fan. But as the speed of the powered fan increases, the speed of air striking the blades of second fan also increases and it starts to rotate. After some time it acquires the same velocity of that of the first fan.

On the same principle the fluid coupling works. In that the impeller act as first fan and the turbine act as second fan. Both impeller and turbine enclosed in an oil tight housing. The impeller is connected the input shaft of the prime mover and the turbine with the output shaft. When the impeller is moved by the prime mover, the fluid in housing experiences centrifugal force and due to curved vanes of the impeller the fluid directed towards the turbine blades. As the fluid strikes the turbine blades it starts rotating. With the increase in the speed of impeller, the velocity of the turbine increases and becomes approximately equal to the impeller speed. The fluid after passing through the turbine blades again return to the impeller.

Working of Fluid Couplings

In a conventional machine construction, transmitting of power from the driver to the driven machine occurs predominantly according to the direct operating principle, for example through shafts, mechanical couplings or gear units.

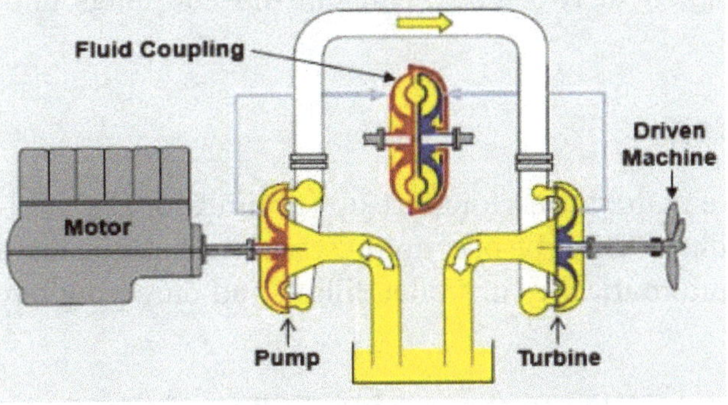

Power Transmission by Indirect Operating Principle

As shown in above figure, power can be also transmitted from a motor (a driver) to a driven machine without direct operating principle using a pump, a turbine and associated piping. It can be seen that the power is transferred from the pump to the turbine without any contact (indirectly). Similarly, in a fluid coupling, transmission of power occurs indirectly based on an indirect operating principle, principle of hydrodynamic power transmission.

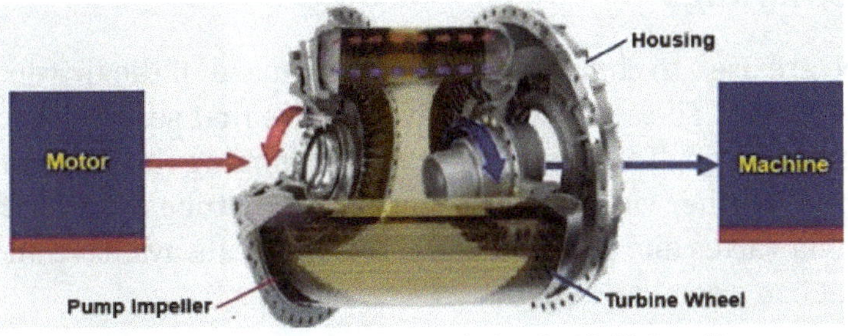

Hydrodynamic Power Transmission in Fluid Coupling

As shown in above figure, in a fluid coupling, pump impeller and turbine wheel are enclosed in a leak proof housing (also called casing or shell) for hydrodynamic power transmission. The pump impeller is connected to the primary driver (motor) and the turbine wheel to the driven machine. The pump impeller transfers the introduced mechanical energy to kinetic energy in fluid flow. The higher energy fluid flows centrifically from the pump impeller to the turbine wheel where a reconversion to mechanical energy (force) takes place. Thus the power is transferred from the pump to the turbine without any contact and thus without wear. Only the bearings and sealing elements are subject to natural wear and tear. Because of the straight radially arranged vanes/blades, the torque is transmitted independently of the direction of rotation and solely by the amount/quantity of oil filling in the coupling.

To generate circulation of the operating fluid for torque/power transmission, a difference in speed is necessary between the pump impeller and turbine wheel. A centrifugal force pressure field is set up that is greater in the faster rotating pump impeller than in the turbine wheel. The difference in speed, usually termed "slip", at the continuous operating point of the coupling is from 2% to 6%, depending on application and coupling size. Immediately after drive motor start-up, slip is 100%, i.e. the pump impeller is driven at the speed of the motor, but the turbine impeller remains stationary.

Types of Fluid Coupling Designs

Fluid couplings are designed in two types: constant-fill couplings and variable-speed (fillcontrolled) couplings.

Constant-fill Couplings

Couplings of this type are mainly used for start-up (to limit torque) and to cushion the torsional vibration of the drive chain. In this type of couplings, various designs mainly differ through adjoining chambers, who are automatically controlled filling and emptying have a significant influence on the start-up behavior.

Constant-fill couplings are sealed to the outside. Filling of the operating fluid in a coupling is carried out before its commissioning. Drive requirements determine the design and filling quantity. The ratio of the operating fluid volume filled to the overall volume of the coupling is called the fill level. The coupling obtains a specific characteristic curve from the fill level. Manually adding or removing operating fluid (at a standstill) enables altering the characteristic curve.

Variable-speed Couplings

Couplings of this type are used to control or regulate the speed of the driven machine over a wide range below the drive speed. These couplings have devices that seamlessly change the transmission behavior during operation. This mainly occurs by changing the fill level. The fill level can be changed during operation either via a radially movable scoop tube or by controlling the operating fluid inlet and outlet via valves and nozzles. These couplings always have an external fluid circuit for filling changes that can also aid cooling.

Construction and Working of Fluid Couplings

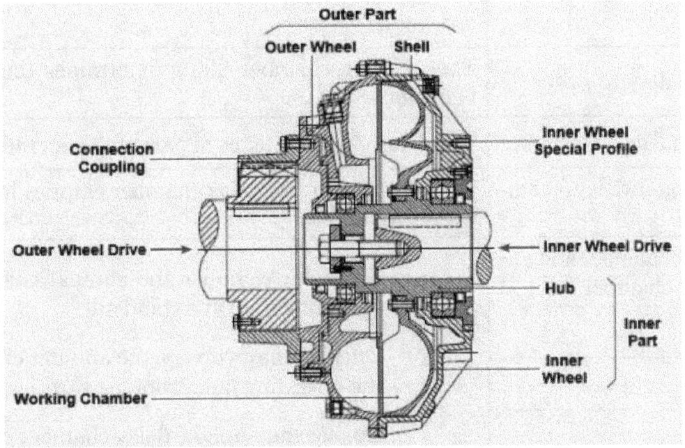

Type T Fluid Coupling with Elastic Connection Coupling

The Voith Turbo fluid coupling Type T (basic design), shown in above figure, consists of a few main components without additional adjoining storage chambers. The shell and the vanned outer wheel forms the working chamber. This working chamber is housed relative to the hub and sealed to the outside. The inner part consists of the hub and the inner (turbine) wheel, which are fixed together.

A fluid coupling is installed between co-axial shafts. Normally a fluid coupling is connected to the shaft of a motor (driving machine). As it is a rigid unit, it is connected to the motor by a flexible connection coupling which absorbs any small assembly misalignment. If necessary or for constructional advantage, a fluid coupling can also be mounted on the motor shaft. In Type T fluid coupling, the full quantity of operating fluid (usually oil) remains in the working chamber (power transmitting chamber) all the time. This type of coupling is suitable for conveyors of less inertia. In general, Type T coupling is suitable for conveyor up to 200 m length, and carrying material of medium bulk density.

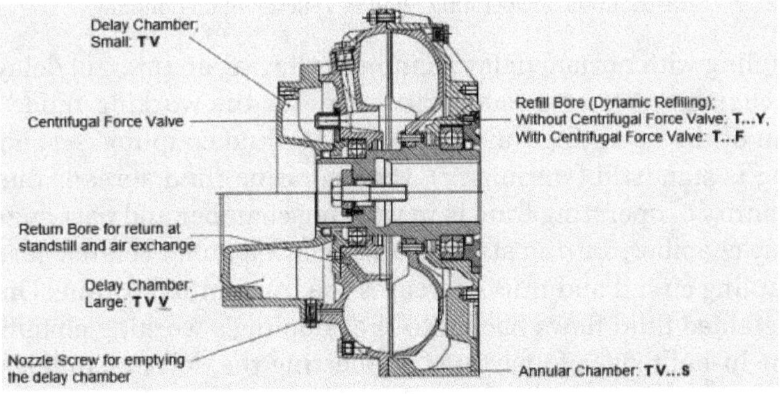

Design Features of Constant-filled Fluid Couplings

The above figure shows design features of constant-filled fluid couplings. They fulfill a variety of requirements for the start-up and operating behavior as needed. The use of adjoining chambers and the volume exchange capability via fixed bores or bores closed with centrifugal force valves enable a great variation of characteristic curves in relation to speed, time and slip.

Voith Type		Design Features	Function Description
Designation			
T		Basic design	The working chamber filling determines the start-up and operating behavior.
TV		Normal delay chamber	The delay chamber takes in part of the operating fluid at a standstill.
TVV		Enlarged delay chamber	During start-up, the delay chamber empties into the working chamber via nozzle bores.
TVS		Delay chamber and annular chamber	• The delay chamber and annular chamber take in part of the operating fluid at a standstill. • During motor run-up, the annular chamber also takes part of the operating fluid from the working chamber. • During start-up, the delay chamber empties into the • working chamber via nozzle bores.

The above table shows description of function for fluid couplings having various design features (for Voith type designation).

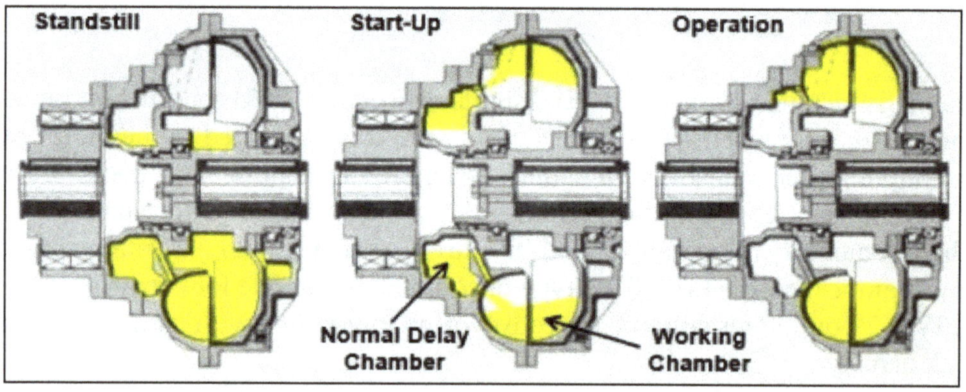

Distribution of Operating Fluid in Type TV Fluid Coupling

Type TV fluid coupling with normal delay chamber is having an integral delay filling chamber. The delay filling chamber acts as a temporary store for the working fluid. The above figure shows distribution of the operating fluid in a Type TV fluid coupling. As shown in the figure, when the coupling is standstill (stationary), the operating fluid spreads due to gravity; with the result part quantity of operating fluid is in working chamber and part quantity of operating fluid is in the delay chamber. During start-up the delay chamber retains part of the operating fluid from the coupling circuit and thus decreases the transmitted torque. Only after run-up is completed, this retained fluid flows back into the coupling's working chamber. The flow rate is governed by the in-built size of apertures connecting the delay chamber with the working chamber. This ensures more (compared to Type V coupling) effective relief for the electric motor during start-up.

In general, Type TV coupling is suitable for conveyor length from 200 to 500 m whereas Type TVV coupling is suitable for conveyor length from 400 to 1000 m (it may be noted that conveyor length is influenced by various factors like bulk density, conveyor inclination, etc.).

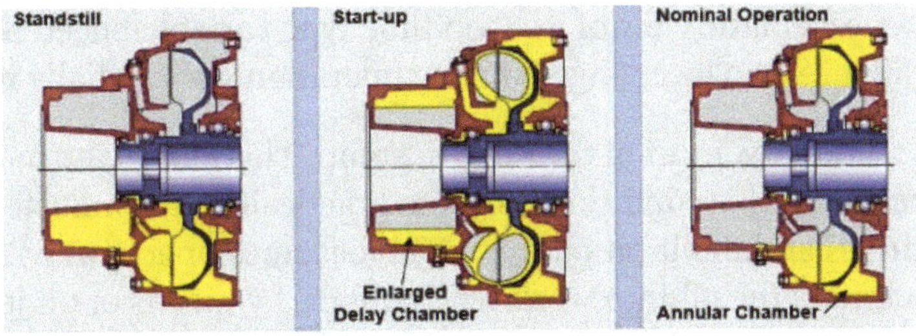

Distribution of Operating Fluid in Type TVVS Fluid Coupling

Type TVVS fluid coupling with enlarged delay chamber is having an enlarged delay chamber and an annular chamber. The above figure shows distribution of the operating fluid in a Type TVVS fluid coupling.

When the motor is started, the portion of operating fluid quantity which is in the working chamber, immediately flows (thrown) into the annular space due to centrifugal action. Thus motor and primary side of the coupling begins the run-up almost without presence of operating fluid in the working chamber. This result into still better performance compared to coupling with enlarged delay chamber.

The machines/conveyor with high inertia can accelerate at ease with this type of coupling. This type of coupling has been used for conveyors up to 2.0 km length. This type of coupling is also used for lesser length high capacity (high power) conveyors when superior performance is needed.

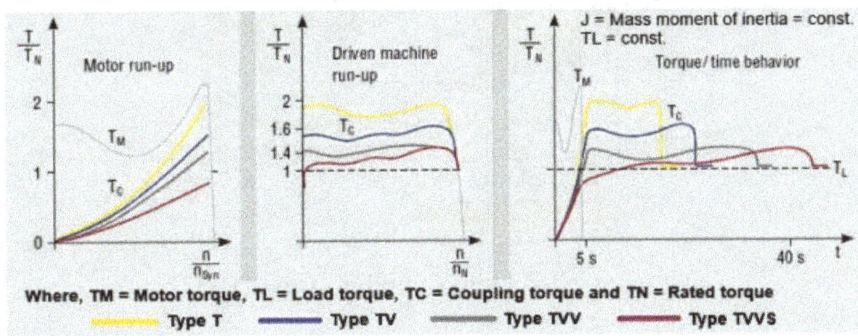

Comparison of Start-up Behaviour of Driven Machine With Squirrel Cage Asynchronous Motor

Comparison of start-up behavior of a driven machine with constant load torque and mass moment of inertia while using various Voith coupling types is shown in above figure. The drive motor is a squirrel cage asynchronous motor.

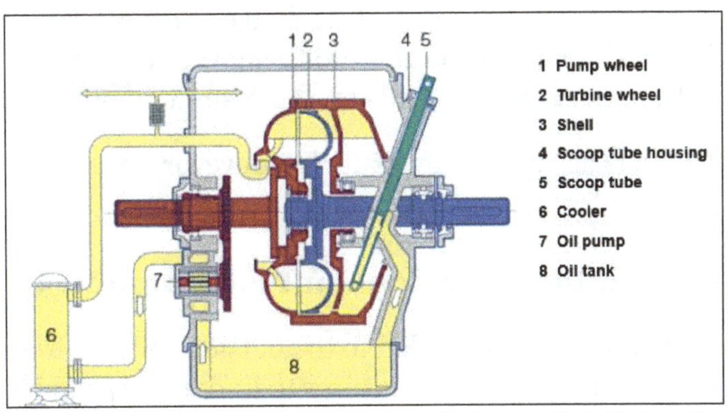

Schematic View of Scoop Tube Type Variable-Speed Fluid Coupling

The above figure shows schematic view of a scoop tube type variable-speed fluid coupling. In this type of coupling, the fill level can be changed during operation via a radially movable scoop tube.

When the coupling is stationary, most of the oil is in sump. This allows the motor to run up practically without load, because in this condition motor is practically not connected to load. The motor side of the coupling drives an inbuilt oil pump, which becomes operative with the rotation of the motor. This oil pump injects the oil into working chamber, by means of oil jet. Subsequently, the quantity of oil in working chamber depends upon the position of scoop tube. The stationary scoop tube mouth intercepts the rotating oil, and thereby empty out the oil which is within the radius formed by scoop tube mouth. Thus, oil quantity in working chamber is governed by scoop tube radial position. During start-up phase, the scoop tube position gets automatically adjusted (retracted), which gradually increases the quantity of oil in working chamber. The gradual increase of oil quantity in working chamber results in optimum starting torque steadily for longer time. This allows the design of a machine (e.g. conveyor) with low starting forces.

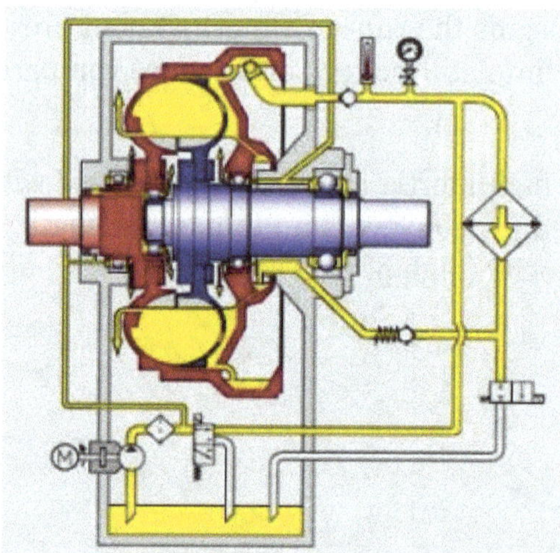

Fill-controlled Fluid Coupling. Type TPKL.

The above figure shows construction of fill-controlled fluid coupling, Type TPKL. This coupling's working principle rests on the metered filling of the rotating working chamber. In this coupling, operating fluid exits the working chamber via nozzles into an annular shell connected to the pump impeller. A dynamic pressure pump (fixed scoop tube) siphons off the operating fluid from this annular shell and then conducts it through a heat exchanger via a catching ring back into the working chamber. This results in an outer, enclosed operating fluid circuit. Working chamber filling changes are carried out indirectly through the addition or removal of operating fluid from this outer circuit. The flow rate of the filling pump depends on the desired fill time of the working chamber.

Variable-speed fluid couplings are used principally in conveyor systems with higher powers and long starting times.

Special Design Considerations for Constant Fill Couplings

As fluid couplings of constant fill type are sealed to the outside and the exchange of gases or fluids with the environment is not easily possible, measures to limit the thermal inner pressure must be implemented.

To generate the operating fluid circulation, a difference in speed, usually termed "slip", is necessary between the pump impeller and turbine wheel. Slip multiplied by the transmitted power represents the power loss of the coupling, which is converted into heat inside the fluid/oil filling. The amount of slip heat generated during the start-up process and in normal operation must be released into the environment via the coupling housing surface to prevent an impermissible temperature rise. The rated coupling output is mainly determined by the power loss which can be dissipated at a still acceptable operating temperature or a reasonable set slip limit. A limitation of the start-up frequency (number of starts per an hour) to reduce thermal stress may also be necessary.

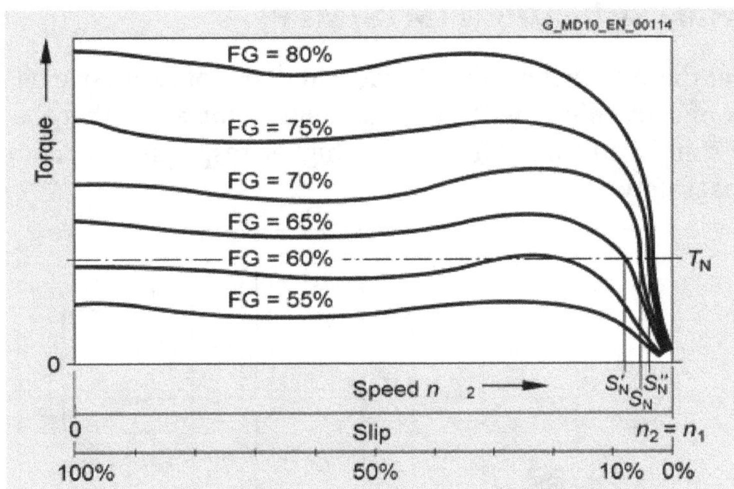

Slip-torque Characteristics for Different Filling Levels FG

As shown in above figure, slip-torque characteristic of a Type T, basic design coupling (without a delay chamber) depends on the quantity of fluid filling (% Fill Level, % FG) in the coupling which enables the transmissible torque on starting up to be via the filling level. It may be noted that with a higher filling level, the starting torque increases, while the operating slip and thus the coupling temperature rise decreases. Conversely, with a lower filling level the starting torque decreases, the coupling becomes softer, while slip and coupling temperature rise. It may be noted that starting torque can be reduced without increasing continuous operating slip by using a coupling with a delay chamber (Type TV).

In general, couplings should not be filled more than 80% of their overall internal volume. Otherwise, the excessive internal pressure produced by heating of the operating fluid may cause serious damage to the coupling and may even lead to fatal injury! In view of this, fill opening should be designed such that the maximum achievable fill level is 80 %. For safety, all constant-fill couplings should also have fusible plugs that release an opening if the thermal stress limit is exceeded. The operating fluid exits from this opening and the power transmission is interrupted.

In addition, mechanical thermal limit switch or thermal measuring unit may be provided.

Fusible Safety Plugs

Fusible plugs (one or more depending on design) protect the coupling from irreparable damage through overheating or overpressure. If a fluid coupling is operated with an impermissibly high slip for a prolonged period, the oil (operating fluid) filling and the coupling housing will overheat. If output of the coupling gets blocked, the input power completely turns into heat in heating up the oil. In such situation, the core of fusible plug's will melt upon reaching a preset temperature and release the oil through this opening's into the environment. This results in stoppage of power transmission through the coupling, separating the input and output drives. Thus fusible plug can prevent the oil from getting too hot and causing serious damages and fire.

For safety, never replace a fusible plug with one having a higher temperature rating than the recommended one or a screw.

Mechanical Thermal Switch/Plug (MTS/MTP)

By adding mechanical thermal switch/plug, leakage and loss of expensive oil through the fusible plug as well as a risk of contaminating the environment in the event that the coupling overheats can be avoided. The standard fusible plug (with a higher response temperature) remains in the coupling for additional safety.

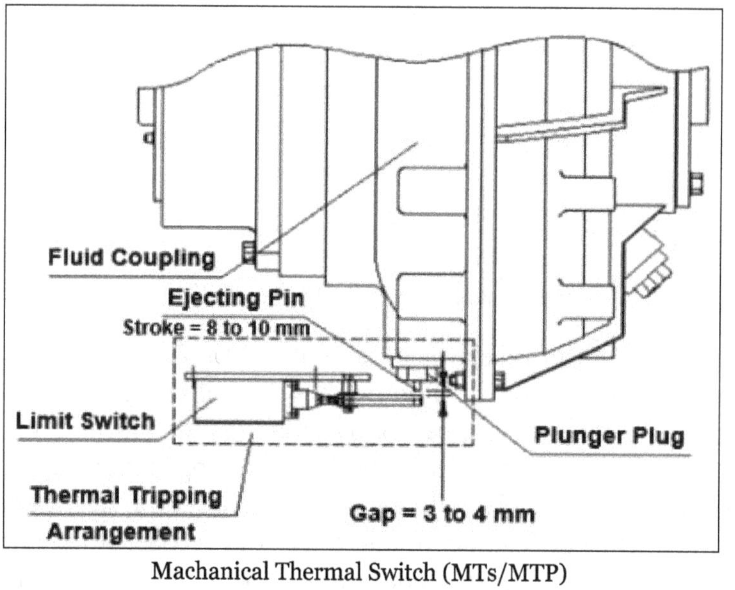

Machanical Thermal Switch (MTs/MTP)

As shown in above figure, mechanical thermal switch consists of a plunger plug screwed on the fluid coupling body (housing). A spring-loaded pin and a chamber filled with solder are integrated in the plunger plug. A cam with limit switch is mounted on stationary base frame. The response/ preset temperature of the plunger plug corresponds to the melting temperature of the solder. Below the response temperature, the solder keeps the pin in its initial position. On reaching the nominal response temperature, the solder releases the pin, and a compression spring presses the pin toward the outside (by 8 to 10 mm) which hits the cam (due to coupling's rotation) and operates the limit switch. The switchgear can cut out the drive motor and/or trigger an optical or acoustic alarm signal. As the housing of the coupling remains closed, no oil will escape. However, once the plunger plug of the mechanical thermal switch has responded, it is no longer usable and needs to be replaced.

However, in case of inner wheel drive (motor connected to turbine wheel) and blocking of driven machine, the mechanical thermal switch does not work.

It is recommended to use mechanical thermal switch and fusible safety plug combination for continuous coupling operation temperatures as under:

Continuous Coupling Operation Temp.	Mechanical Thermal Switch	Fusible Safety Plug
≤ 85 °C	110 °C	140 °C
> 85 °C to 110 °C	140 °C	160 °C

Non-contacting Temperature Monitors

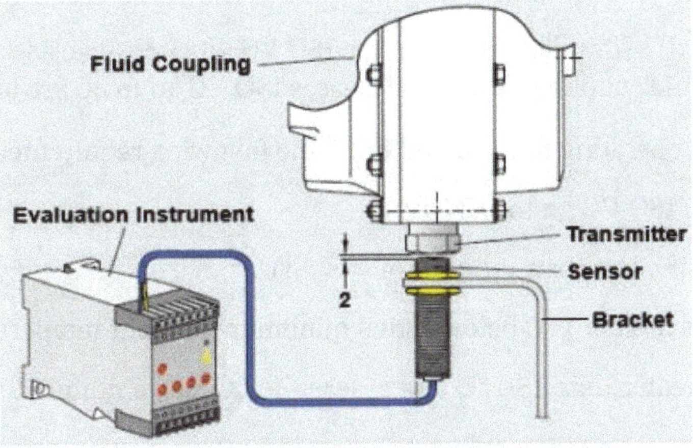

Non-contacting Temperature Monitoring system.

Above figure shows typical non-contacting temperature monitoring system (for example, EOC system by Siemens/A. Friedrich Flender AG). By adding non-contacting temperature monitor with display and relay output, leakage and loss of oil through the fusible plug in the event that the coupling overheats can be avoided. After coupling has cooled down, the system resets. The standard fusible plug (with a higher response temperature) remains in the coupling for additional safety.

To initiate desired action on the coupling reaching response/preset temperature, noncontacting temperature monitors are available in two types:

1. Non-contacting thermal switch unit (e.g. BTS by Voith).

2. Non-contacting thermal measuring device (e.g. BTM by Voith).

Suggested temperature settings for them are as under:

Set Point I (Alarm)	Set Point II (Trip)	Fusible Safety Plug
100 °C	110 °C	140 °C
125 °C	135 °C	160 °C

The thermal switching element (e.g. BTS) must not be used as safety device to limit the maximum permissible temperature of the fluid coupling in potentially explosive atmospheres.

Operating Fluids

In fluid couplings, fluid flow transfers the power from the pump impeller to the turbine runner.

Hence the operating fluid is an important design element for the whole system. The density and viscosity of the fluid are of primary importance. Higher the density of the operating fluid, better the transmission capacity. Lower the viscosity lower the speed losses of the circumferential mass flow in the flow channel. Other important properties are high viscosity index, high specific heat and high operating temperature as another key task is to take in the heat resulting from the power loss and then transfer it to the coupling components.

The operating fluid is selected on the basis of the parameter requirements for the power transmission, the material compatibility and the minimum lubricant capability for antifriction bearings.

HLP (mineral-oil based) hydraulic oils within the ISO VG 32 viscosity class are the standard operating fluids used for fluid couplings. For special cases ISO VG 10 to 46 are usable.

The mineral-oil based operating fluid should fulfill the following requirements.

- Viscosity class: ISO VG 32 to DIN 51519.
- Starting viscosity: less than 15000 mm^2s^{-1} (cSt).
- Pour point: the limit is 4 °C below actual minimum ambient temperature or lower.
- Flash point: greater than 180 °C and at least 40 °C above nominal response temperature of fusible plug.
- Fire point (for couplings used in hazardous areas): at least 50°C above max. surface temperature
- Resistance to aging: aging-resistant refined product
- Compatibility with seals: NBR (Nitril-Butadien rubber) and FPM/FKM (Viton rubber).
- Cleanliness grade: 21/19/16 as per ISO 4406 (Minimum requirement).

Many fluids including hydraulic oils HLP 32 to DIN 51524, Part 2; lube oil CLP 32 to DIN 51517, Part 3; steam turbine oils LTD 32 to DIN 51515, Part 1; HD engine oils SAE 10W; ATF type A Suffix A (TASA) and type Dexron II meet above requirements and can be used.

Some proposed operating fluids (mineral oils) are as under.

Manufacturer	Designation	Pour Point °C	Flash Point °C
BP	Energol HLP-HM 32	−30	216
Castrol	Hyspin SP 32	−28	200
	Hyspin AWS 32	−27	200
ExxonMobil	Nuto H 32	−24	212
	DTE 24	−27	220
	Mobil SHC 524	−54	234
Kuwait Petroleum	Q8 Haydn 32	−30	208
	Q8 Holst 32	−30	208

Shell	Tegula V 32	-33	211
Total	Azolla ZS 32	-27	210
HP (Hindustan Petroleum)	HP ENCLO 32	-3	190
HP (Hindustan Petroleum)	HP ENCLO VT 32	-37	224
IOC (Indian Oil Corporation)	IOC - SERVOSYSTEM 32	-6	190
IOC (Indian Oil Corporation)	IOC - SERVOSYSTEM HLP 32	-21	200

Hydraulic Torque Converter

A torque converter is a type of fluid coupling which is used to transfer rotating power from the engine of a vehicle to the transmission. It takes place of mechanical clutch in an automatic transmission. The main function of it is to allow the load to be isolated from the main power source. It sits in between the engine and transmission. It has the same function as the clutch in manual transmission. As the clutch separates the engine from the load when it stops, in the same way it also isolates the engine from load and keep engine running when vehicle stops. Its main functions are:

1. It transfers the power from engine to the transmission input shaft.
2. It drives the front pump of the transmission.
3. It isolates the engine from the load when the vehicle is stationary.
4. It multiplies the torque of the engine and transmits it to the transmission. It almost doubles the output torque.

Working Principle

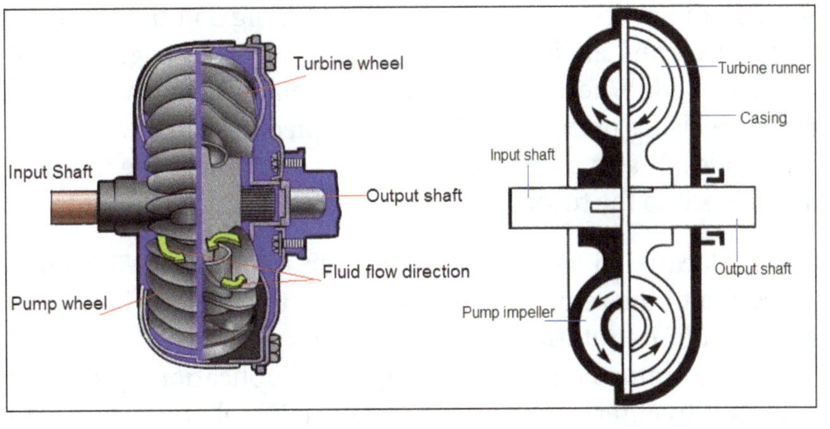

For understanding the working principle of torque converter, let's take two fans. One fan is connected to the power source and other is not connected with power source. When first fan connected to the power source starts moving, the air from it flows to the second fan which is stationary. The air from the first fan strikes on the blades of the second fan and it also starts rotating almost at the same speed to the first one. When the second fan is stopped, it does not stop the first one. The first fan keeps rotating.

On the same principle the torque converter works. In that the impeller or pump acts as first fan which is connected to the engine and turbine act as the second fan which is connected to the transmission system. When the engine runs, it rotates the impeller and due to the centrifugal force the oil inside the torque converter assembly directed towards the turbine. As it hits turbine blades, the turbine starts rotating. This makes the transmission system rotate and the wheels of the vehicle moves. When engine stops, the turbine also stops rotating but the impeller connected the engine keeps moving and this prevent the killing of engine.

Main Parts

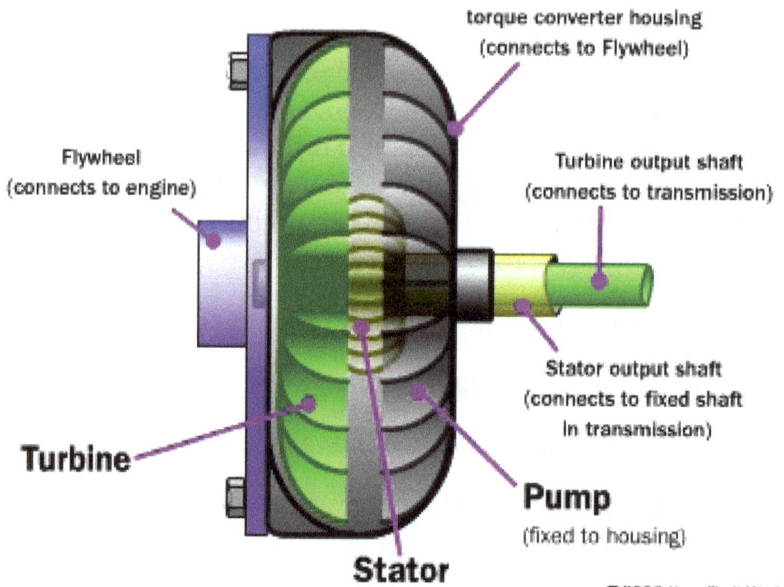

The torque converter has three main parts

- Impeller or Pump: The impeller is connected to the housing and the housing connected to the engine shaft. It has curved and angled vanes. It rotates with the engine speed and consists of automatic transmission fluid. When it rotates with the engine, the centrifugal force makes the fluid move outward. The blades of the impeller are designed in such a way that it directs the fluid towards the turbine blades. It acts as centrifugal pump which sucks the fluid from the automatic transmission and delivers it to the turbine.

- Stator: The stator is located in between the impeller and turbine. The main function of the stator is to give direction to the returning fluid from the turbine, so that the fluid enters to the impeller in the direction of its rotation. As the fluid enters in the direction of the impeller, it multiplies the torque. So stator helps in the torque multiplication by changing the direction of the fluid and allows it to enter in the direction of the impeller rotation. The stator changes the direction of fluid almost upto 90 degree. The stator is mounted with a one way clutch that allows rotating it in one direction and preventing its rotation in other direction. Turbine is connected to the transmission system of the vehicle. And the stator is placed in between the impeller and turbine.

• Turbine: Turbine is connected to the input shaft of the automatic transmission. It is present at the engine side. It also consists of curved and angled blades. The blades of the turbine are designed in such a way that it can change the direction of the fluid completely that strikes on its blades. It is the change in the direction of the fluid that forces the blades to move in the direction of the impeller. As the turbine rotates the input shaft of the transmission also rotates and made the vehicle to move. The turbine is also has a lock up clutch at its back. The lock up clutch comes into play when the torque converter achieves coupling point. the lockup eliminates the loses and improves the efficiency of the converter.

Working of Torque Converter

It has three stages of operations:

• Stall: During stall (stop) condition of the vehicle, the engine is applying power to the impeller but the turbine cannot rotate. This happens, when the vehicle is stationary and driver has kept his foot on the brake paddle to prevent it from moving. During this condition maximum multiplication of torque takes place. As the driver removes its foot from the brake paddle and presses the accelerator paddle, the impeller starts moving faster and this set the turbine to move. At this situation, there is a larger difference between the pump and turbine speed. The impeller speed is much greater than the turbine speed.

• Acceleration: During acceleration, the turbine speed keeps on increasing, but still there is large difference between the impeller and turbine speed. As the speed of the turbine increases the torque multiplication reduces. During acceleration of the vehicle the torque multiplication is less than that is achieved during stall condition.

• Coupling: It is a situation when the turbine achieved approximately 90 percent speed of the impeller and this point is called coupling point. The torque multiplication seizes and becomes zero and the torque converter behaves just like a simple fluid coupling. At the coupling point the lock up clutch come into play and locks the turbine to the impeller of the converter. This puts the turbine and impeller to move with the same speed. Lock up clutch engages only when coupling point is achieved. During coupling the stator also starts to rotate in the direction of the impeller and turbine rotation.

- The maximum torque multiplication takes place during stalling condition.
- The stator remains stationary before coupling point and helps in the torque multiplication. As the coupling attained, stator stops torque multiplication and starts rotating with the impeller and turbine.
- Lock up clutch engages when coupling point is achieved and removes the power losses resulting in increased efficiency.

Advantages

- It produces the maximum torque as compared with the vehicle equipped with clutch.
- It removes the clutch pedal.
- It makes the job of driving a vehicle easier.

Disadvantages

- Its fuel efficiency is low as compared with the vehicle with manual transmission.

Application

- The torque converter is used in the vehicle that is equipped with the automatic transmission. It is also used in industrial power transmission such as conveyer drives, winches, drilling rigs, almost all modern forklifts, construction equipment, and railway locomotives.
- It is used in marine propulsion systems.

References

- TechZone-HydraulicPumps: hydraulicspneumatics.com, Retrieved 18 July, 2019
- Fluid-coupling: mechanicalbooster.com, Retrieved 14 January, 2019
- Construction-Working-Operation-and-Maintenance-of-Fluid-Couplings: practicalmaintenance.net, Retrieved 21 June, 2019
- Torque-converter: mechanicalbooster.com, Retrieved 30 August, 2019

5

Machine Design and Machine Tools

The process of converting resources or energy into useful mechanical forms in order to obtain desired output is known as machine design. The machines which are used to shape metals and other rigid materials are known as machine tools. This chapter has been carefully written to provide an easy understanding of the varied facets of machine design and machine tools as well as the manufacturing processes associated with them.

Machine Design

Machine design is the process of engineering design. A machine is made up of mechanisms that work together to satisfy the requirements of what the machine needs to accomplish. Mechanisms can still be used independently, such as an umbrella, but cannot accomplish what a machine can because a machine controls energy in addition to motion. Machine design takes into account kinetics and kinematics, which deal with motion and the forces on an object in motion. Machine design is applied through a specific process including determining what the machine needs to do, benchmarking and defining goals and requirements, brainstorming, evaluating and selecting from the different options, creating an in-depth design, creating and testing a prototype, and finally manufacturing the machine.

The concept of machine design may be illustrated in the figure below:

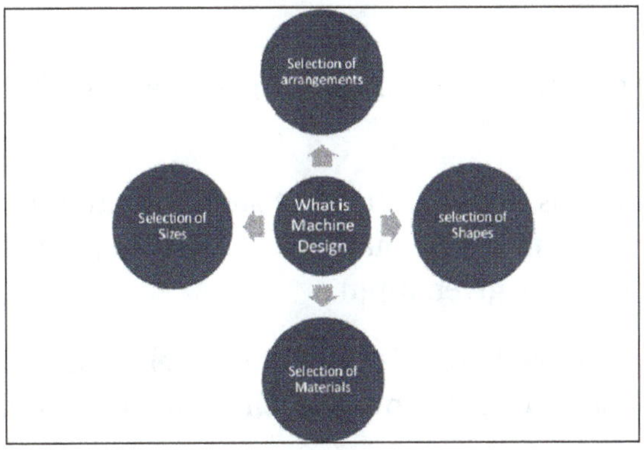

Concept of Machine Design

For example, the process of design of a belt drive consists of:

- Selection of arrangement of mechanical elements like pulleys, belt, shaft, keys bearings etc.
- Selection of shapes of these elements.
- Selection of materials of these mechanical elements.
- Selection of sizes of these mechanical elements.

Most of the problems in mechanical engineering design or specifically in machine design, do not have a unique right answer. There are nearly endless number of workable designs, none of which could be called an incorrect answer. But of the correct answers, some are obviously better than others.

Design Procedure

Design Procedure

The general procedure that is followed in machine design is given in the above figure. It consists of following steps:

1. Definition of Problem: Define the design problem giving all input parameters, output parameters and constraints.

2. Synthesis: Once the problem is defined, the next step is synthesis. Synthesis is the process of selecting or creating the mechanism for the machine and the shapes of the mechanical elements so as to get the desired output with the given input.

3. Analysis of Forces: Draw the free body diagram of each element of the machines. Find out the forces (including moments and torque) acting on each element by force analysis.

4. Selection of Material: Select the suitable material for each element. Four basic factors that are to be considered while selecting the materials are: availability, cost, mechanical properties, and manufacturing considerations.

5. Determination of Mode of Failure: Before finding out the dimensions of the elements, it is necessary to know the type of failure by which the element will fail when put into the use.

6. Selection of Factor of Safety: Based on the application, select the factor of safety. Knowing factor of safety and material strength, determine the permissible or design stress.

7. Determination of Dimensions: Find the dimensions of each element of the machine by considering the forces acting on the element and the permissible stress.

8. Modification of Dimensions: Modify the dimensions of the elements on the higher side, if required, based on the following considerations:

- Selection of standard parts available in the market.
- Convenience of assembly.
- Convenience of Manufacturing.

9. Preparation of Drawings: Prepare working drawing of each element or component with minimum two views showing following details:

- Dimensions.
- Dimensional Tolerances.
- Surface Finish.
- Geometrical Tolerances.
- Special production requirement like heat treatment.

Prepare assembly drawing giving part number, overall dimensions and part list. The component drawing is supplied to the shop floor for manufacturing purpose, while assembly drawing is supplied to the assembly shop.

10. Preparation of Design Report: Prepare design report containing details about step 1 to step 8.

Design Considerations

Design considerations are the characteristics which influence the design of the element or perhaps, the entire system. Normally, a number of such characteristics have to be considered in any design problem.

In a given design problem, the design engineer should identify the various design considerations and incorporate them in the design process in their order of importance. For example, in the design of a spring, two most significant design considerations are strength and stiffness.

Some of the common design considerations are as follows:

1. Strength: The machine elements are subjected to any one or combination of loads like forces, bending moments and torque

A machine element should have sufficient strength to avoid failure either due to yielding or due to fracture under the loads.

2. Rigidity: A machine element should have sufficient rigidity so that its linear as well as angular deflections, under the loading are within permissible limits.

3. Reliability: The reliability is defined as the probability that a component, system or device will

perform without failure for a specific period of time under the specified operating conditions. A machine element should have reasonable good reliability so that it can perform its function satisfactorily over its life span.

4. Safety: A machine element should be designed such that it ensures safety of users and machine.

5. Cost: The life cycle cost of the machine element consists of production cost, operating cost, maintenance cost and disposal cost. A machine element should have a minimum possible life cycle cost.

6. Weight: A machine element should have a minimum possible weight.

7. Ergonomics: Ergonomics is defined as the scientific study of the man-machine working environment relationship and the application of anatomical, physiological, and psychological principles to solve the problems arising from the relationship.

The objective of ergonomics is to make the machine fit for user rather than to make the user adopt himself or herself to the machine. If the user is likely to communicate directly with the machine element, it should be designed with ergonomic considerations.

8. Aesthetics: Aesthetics deals with the appearance of the product. In a present day of buyer's market, with a number of products available in the market are having most of the parameters identical, the appearance of the product is often a major factor in attracting the customer. This is particularly true for consumer durables like automobiles, domestic refrigerators, television, mobile etc.

9. Manufacturing: In a design of machine element, the selection of manufacturing processes must be given a due importance. The manufacturing processes should be selected such that the machine element can be produces with minimum manufacturing cost and as far as possible, with existing manufacturing facilities.

10. Conformance to Standards: A design of machine element should conform to the national and international standards and codes.

11. Assembly: A machine element or a product should be designed such that it facilitates to minimize the assembly cost and time.

12. Friction and Wear: Friction and wear are major contributing factors for reducing the life of machine elements and increasing the power loss. The friction can be reduced by improving the surface finish, adequately lubricating the surfaces and replacing the sliding motion by rolling motion. The wear can be reduced by increasing the surface hardness.

13. Life: A machine element should be designed for an adequate life.

14. Vibrations: A machine element should be designed so as to keep the vibrations at minimum level.

15. Thermal considerations: A machine element should be able to withstand the temperature to which it may be subjected. In addition, it should dissipate the heat generated, if any.

16. Lubrication: In the design of machine elements, due considerations must be given for the lubrication of the elements, if there is relative sliding or rolling motion between the elements.

17. Maintenance: A machine element should be such that it can be easily repaired or serviced.

18. Flexibility: A machine element should be flexible so that the modifications can be carried out with minimum efforts.

19. Size and Shape: As far as possible, standard sizes and shaped should be adopted for machine elements.

20. Stiffness: Whenever stiffness is a functional requirement like in springs, the machine element should be designed with a precise value of required stiffness.

21. Corrosionl: A machine element should be corrosion resistant. This can be achieved by a proper selection of material and adopting the surface coating.

22. Noise: The machine element should be designed such that the noise during operation is at minimum.

Manufacturing Processes

Manufacturing process is basically a complex activity, concerned with people who have a broad number of disciplines and expertise and a wide range of machinery, tools, and equipment with numerous levels of automation, such as computers, robots, and other equipment. Manufacturing pursuits must be receptive to several needs and developments.

Beside above, all the future technicians must understand the basic needs of workshop routines in terms of man, equipment, material, methods, revenue and other infrastructure conveniences needed to be placed properly for maximum shop or plant layouts and other support solutions effectively regulated or positioned in the field or industry within a properly planned manufacturing firm.

Casting

Casting manufacturing is a process in which liquefied material, such as molten metal, is poured into the cavity of a specially designed mold and allowed to harden. After solidification, the workpiece is removed from the die to undergo various finishing treatments or for use as a final product. Casting methods are typically used to create intricate solid shapes, and cast products are found in a wide range of applications, including automotive components, aerospace parts, electronics, mechanical devices, and construction supplies.

Different Types of Casting and the Casting Process

Although casting is one of the oldest known manufacturing techniques, modern advances in casting technology have led to a broad array of specialized casting methods. Hot forming processes, such as die casting, investment casting, plaster casting, and sand casting, each provide their own unique fabrication benefits. Comparing both the advantages and disadvantages of the common types of casting processes can help in selecting the method best-suited for a given production run.

Sand Casting

Sand casting typically relies on silica-based materials, such as synthetic or naturally-bonded sand. Casting sand generally consists of finely ground, spherical grains that can be tightly packed together into a smooth molding surface. The casting is designed to reduce the potential for tearing, cracking, or other flaws by allowing a moderate degree of flexibility and shrinkage during the cooling phase of the process. The sand can also be strengthened with the addition of clay, which helps the particles bond more closely. Many automotive products, such as engine blocks and housings, are manufactured through sand casting.

Advantages of Sand Casting

Sand casting's main advantages as a casting process include:

- Relatively inexpensive production costs, especially in low-volume runs.
- The ability to fabricate large components.
- A capacity for casting both ferrous and non-ferrous materials.
- A low cost for post-casting tooling.

Disadvantages of Sand Casting

Despite its benefits, sand casting yields a lower degree of accuracy than do alternate methods and it can be difficult to sand cast components with predetermined size and weight specifications. Furthermore, this process has a tendency to yield products with a comparatively rough surface finish.

Investment Casting

Investment casting uses a disposable wax pattern for each cast part. The wax is either injected directly into the mold or pre-coated with a liquid refractory material before injection. Molten forming material is then poured into the mold and allowed to harden and set into the shape of the wax pattern. The component is then ejected, while the wax pattern is melted out of the casting and made available for reuse. Investment casting is often used to manufacture parts for the aerospace and automotive industries, as well the military. Some of the central advantages and disadvantages of investment casting include:

- A high degree of accuracy and precise dimensional results.
- The ability to create thin-walled parts with complex geometries.
- The capacity for casting both ferrous and non-ferrous materials.
- Relatively high-quality surface finish and detail in final components.

Although it is highly precise, investment casting is usually more expensive than other comparable casting techniques, and is typically only cost-efficient when sand or plaster castings cannot be used. However, the expense can sometimes be compensated for with reduced machining and tooling costs due to investment castings' quality surface results.

Plaster Casting

Plaster casting is similar to the sand casting process, using a mixture of gypsum, strengthening compound, and water in place of the sand. The plaster pattern is typically coated with an anti-adhesive compound to prevent it from becoming stuck against the mold and the plaster is capable of filling in any gaps around the mold. Once the plaster material has been used to cast a part, it usually cracks or forms defects, requiring it to be replaced with fresh material. The advantages offered by plaster casting include:

- A very smooth surface finish.
- The ability to cast complex shapes with thin walls.
- The capacity for forming large parts with less expense than other processes, such as investment casting.
- A higher degree of dimensional accuracy than that of sand casting.

This process tends to be more expensive than most sand casting operations, and may require frequent replacements of plaster molding material. It is usually more effective and cost-efficient when the quality of the surface finish is an important requirement.

Die Casting (Metal Casting Process)

Die casting is a method of molding materials under high pressure and usually involves non-ferrous metals and alloys, such as zinc, tin, copper, and aluminum. The mold is coated with lubricant to help regulate the die's temperature and to assist with component ejection. Molten metal is then injected into the die under high pressure, which remains continuous until the workpiece solidifies. This pressurized insertion is rapid, preventing any segment of the material from hardening before being cast. After the process is completed, the component is taken out of the die and any scrap material is removed. A few of the major advantages provided by die casting include:

- Close size and shape tolerances.
- High component dimensional consistency and uniform design.
- A reduced need for post-casting machining.

Despite its advantages, die casting as a metal casting process has relatively high tool costs, making it more cost-efficient in high-volume product runs. It can also be difficult to ensure the mechanical properties of a die cast component, meaning these products usually do not function as structural parts.

Forming

Metal forming is the approach of creating the metallic components by deforming the metal but not by removing, cutting, shredding or breaking any part. Bending, spinning, drawing, and stretching are a few important metals forming process in manufacturing. The metal press such as die and punching tools are implemented for this manufacturing process.

Advantages: Same equipment can be utilized for manufacturing various components by simply changing the dies.

Disadvantages: High apparatus and tooling expenses. Heat treatment must be applied afterwards.

Forming processes involve applying forces to the material being shaped. A good way of telling how a given material responds to applied force is to look at diagrams representing its stress-strain behaviour. Figure shows the stress-strain curves, at room temperature, for two different metals. The two important things for the feasibility of squeezing-type processes are the point at which the solid starts to flow and the extent to which it can be persuaded to flow before it separates (i.e. fails). This is described by two properties, yield stress (or flow stress) and ductility. Remember that the yield stress is a good measure of the strength of a ductile material.

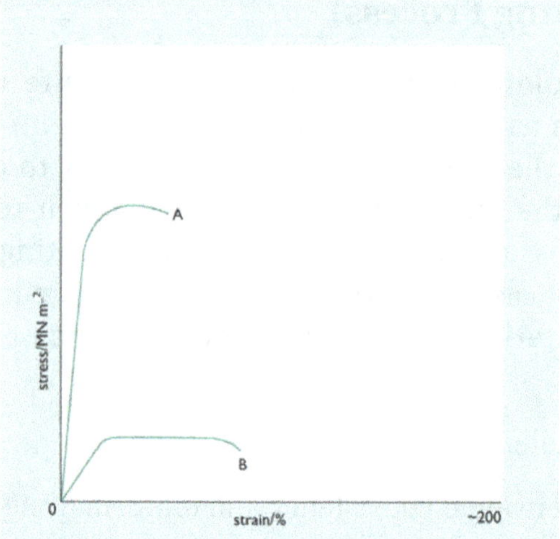

Schematic stress-strain curves for: A – a steel, B – a ductile metal such as lead.

If you look at the curves, you can see that the one for the steel (curve A), after the elastic region, shows plastic deformation up to a strain of about 40 per cent. This provides a measure of the ductility of the steel, and the extent to which it can be squeezed, stretched or bent at this temperature. Curve B (for lead) shows much higher ductility and a much lower yield stress.

However, 40 per cent is not a lot of strain for many manufacturing purposes. Being able to change a material's dimensions by only 40 per cent would mean that forming would be virtually a waste of time. In addition, once the steel in Figure has been strained plastically, it has also increased in yield stress and become harder.

The steel can be loaded elastically up to its yield strength, point A in Figure. Removing the load below this level leaves no permanent extension, i.e. the steel can return to its original size and shape. If the material is taken above its yield strength, say to point B, then even after unloading, the steel is permanently deformed. What's more, when the material is then reloaded, it has to be loaded to a stress equal to point B – i.e. more than its original yield point at A – before it continues to flow. This is called work hardening: the plastic deformation causes an increase in strength (and hardness). But further plasticity is then limited. If the material is loaded to point C, where about half the available plasticity has been 'used up', then on reloading only the remaining plasticity from C to D is available before the material breaks.

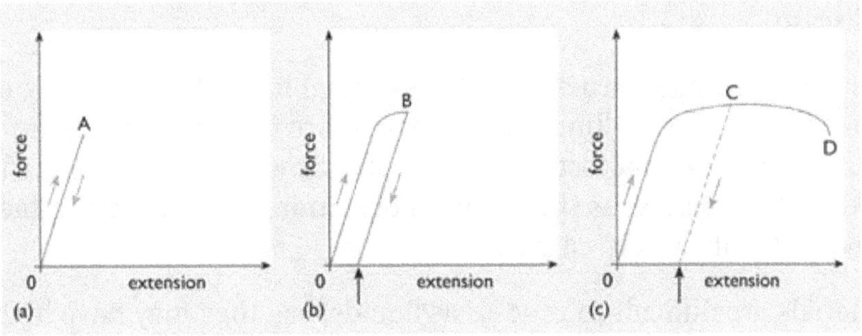

The effect of progressively straining a material.

However, all is not lost. The work hardening effect can be eliminated, and the original, softer, condition restored, by annealing the metal. This involves heating to a temperature where the atoms in the material become more mobile, so that the material softens. This type of process is widely applied in manufacturing whenever a part-formed product has been worked so much that it is in danger of cracking; and is why blacksmiths keep reheating a horseshoe during working. The precise mechanism of softening varies from material to material but it occurs for most materials at a high homologous temperature. Homologous temperature, T_H, is the ratio of the operating temperature of a material to its melting point (in kelvins, K).

In the context of forming, if a metal is 'hot worked', it is deformed at a temperature where there is virtually no work hardening, and this resoftening effect carries on continuously. Very high strains can be imposed during the forming process.

Conversely 'cold working' does result in work hardening, so that the material gets harder the more it is deformed. 'Cold working' does not literally mean cold, though it is all measured relative to the homologous temperature. The rule of thumb usually employed is that cold forming occurs at homologous temperatures below 0.3 and hot forming occurs at homologous temperatures above 0.6. In between the two, there is a region known as warm forming. This means that if tungsten is worked at 1000 °C (1273K) it is being cold formed, as its melting point is 3410 °C (3683K). Conversely, at room temperature, solder can be hot worked! So working temperature is all about being able to deform the metal without failure.

Forming vs. Casting

As the stresses needed to make solids flow are considerably higher than those required for liquids, forming processes normally require a lot of energy and strong, resilient tooling. This means high expenditure on capital equipment as well as tooling and energy. As a result, forming is often economically viable only for production volumes large enough to justify the high tooling costs.

So when do we use forming rather than casting? There are three reasons why, for many products, forming is preferable to casting.

1. Geometry: Products with one dimension significantly different in size from the others are most suitable for forming processes – 'long' products such as rails or 'thin' products such as car-body panels are usually made by a forming process. Imagine trying to cast a 50 metre length of pipe; large forces would be needed to squeeze the metal down into the mould, and it might be difficult to keep the metal liquid for long enough. There would also be a lot of scrap to discard.

2. Microstructure: the microstructure of the material has a direct bearing on the properties of the final product. Controlling the microstructure is easier during forming than during casting. Also, the type of microstructures produced by forming are inherently stronger than those produced by casting, as the products of forming do not contain the dendritic structure and porosity inherent to castings.

3. Some materials are difficult to process as liquids, e.g. they may have high melting points, or react with the atmosphere.

Forming Processes

Forming processes are used to convert cast ingots into basic product forms such as sheets, rods and plates. However, here we will concentrate on forming processes that produce end products or components. There are some basic shapes that lend themselves to manufacture by forming. Forming processes are particularly good at manufacturing 'linear' objects, that is, long thin ones, where the product has a constant cross section. Forming processes involve moving the material through an opening with the desired shape. These processes are used for making components such as fibres, wires, tubes and products such as curtain rails. The plastic ink tube in your ballpoint pen was almost certainly produced by this method.

Extrusion

The principle of this process is very similar to squeezing toothpaste from a tube. Material is forced through a shaped hollow die in such a way that it is plastically deformed and takes up the shape of the die. The hole in the die can have almost any shape, so if the die is circular, for example, a wire or rod is produced.

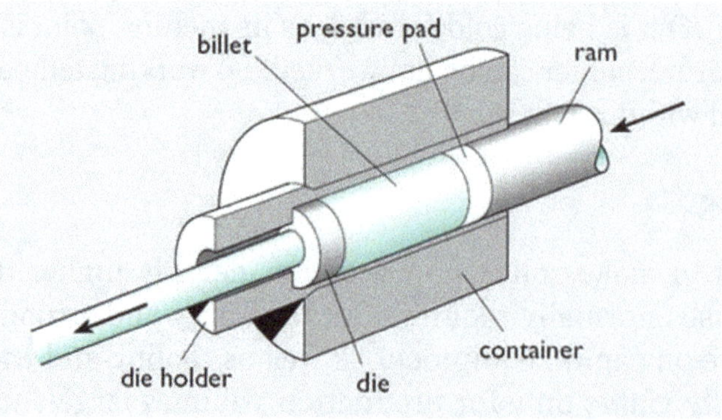

Metal extrusion

It is also possible to produce hollow sections using extrusion. In this case, the die contains a short piece (or mandrel) in the shape of the hole. This mandrel is attached to the die by one or more 'bridges'. As the extruded material encounters the bridges it is forced to separate, but it flows around the bridges and joins up again, much the same as water flowing around the piers of a bridge. Figure shows such a 'bridge die'. This works successfully even for processing of solid metals.

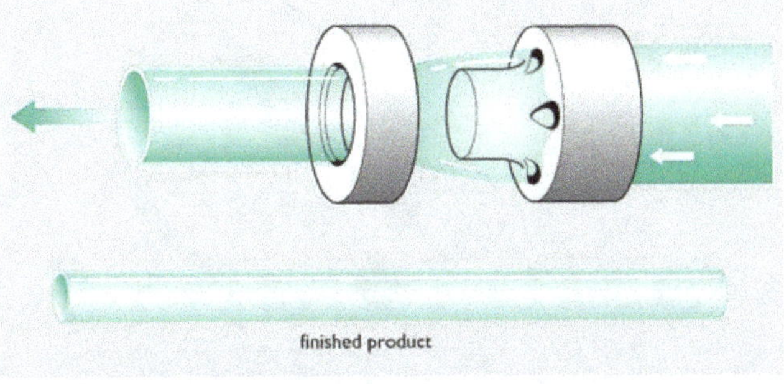

Extrusion bridge die making a hollow section product. Note that in the picture the die has been split to show the material passing through it. In reality, the die and the ring fit together, with a gap for the extruded material to flow through.

Extrusion can be used on most materials that can plastically flow as a solid, and solid metals and alloys are frequently extruded. To reduce the stresses required, and therefore the size and cost of the extrusion machine, and also to ensure hot working conditions, a metal is usually extruded at a high homologous temperature, usually between 0.65 and 0.9. This allows large changes in the shape of the material and hence large strains without fracture. During metal extrusion the raw material in the form of a metal ingot, known as the billet, is heated and pushed through the die by a simple sliding piston or ram.

The mechanism for extruding thermoplastics is illustrated in Figure. In this case a rotating screw is used to transfer the raw material in the form of granules through a heated cylinder to the die, just like in the case of injection moulding for polymers. The thermoplastic granules are compressed and mixed by the screw (the granules may contain a second constituent such as colouring). The material softens and melts as the temperature rises due to heating through the walls of the cylinder, and also from the heat generated within the thermoplastic as it is sheared by the screw. The thermoplastic flows through the die and emerges with a constant cross section in the shape of the die aperture. An almost infinite variety of cross-sectional shapes can be produced.

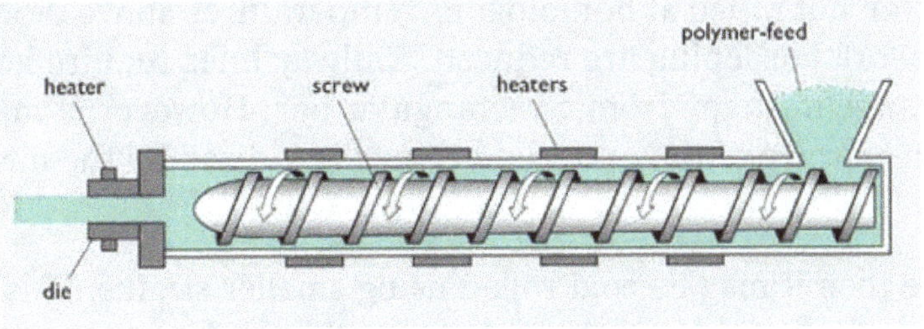

Thermoplastics extrusion

Rolling

In rolling, material is passed through the gap between two rotating rollers that squeeze the material as it passes between them. The rolled material emerges with a thickness roughly equal to the gap between the rollers. When the rollers are cylindrical, rolling produces material in the form of plate or sheet. Sheet steel and aluminium for the bodies of cars and domestic appliances is made this way. Rolled sheet is often termed a 'semifinished' product, as it requires further processing to shape it into the final product.

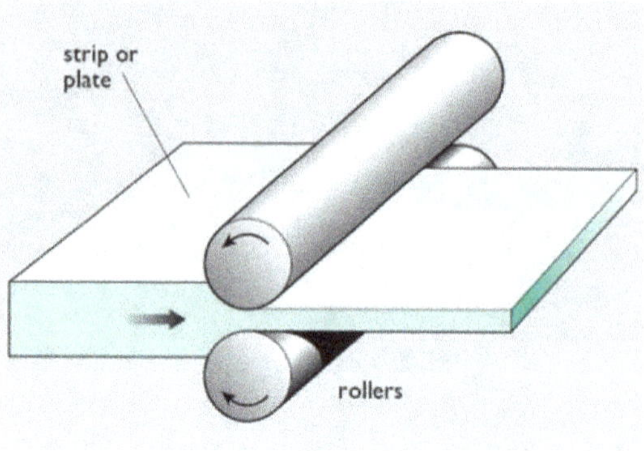

Rolling

Rolling is not restricted to flat sheets, though. If the desired product has a contoured surface, then by using profiled rollers the contour can be rolled on. If the surface pattern needs to be deeper than is possible during one rolling pass then multiple rollers can be used; for example, railway tracks are made by rolling between pairs of progressively deeper contoured rollers. The various stages for rails are shown in Figure.

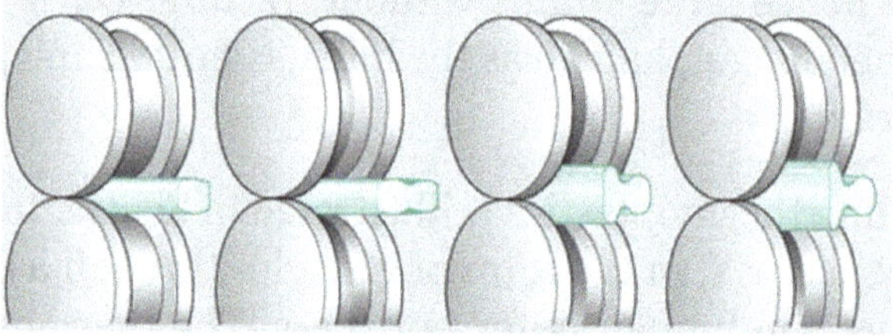

Stages in rolling railway track

In common with other forming processes, metals may be hot or cold rolled. The significant differences between hot and cold rolling are in the amount of energy needed to roll a given volume of material and in the resulting microstructures. The cooler the metal, the higher its yield stress and the more energy has to be supplied in order to shape it. As in extrusion, metals in large lumps are often hot rolled at homologous temperatures above 0.6. At this temperature the yield stress and work hardening are reduced. Railway lines require hot rolling in order to achieve the large change in shape from a rectangular bar. However, a major disadvantage of hot rolling is that the surface of the material becomes oxidised by the air, resulting in a poor surface finish.

If the metal is ductile then it may be cold rolled using smaller strains. This has some advantages: the work hardening at these temperatures can give the product a useful increase in strength. During cold rolling, oxidation is reduced and a good surface finish can be produced by using polished rollers. So, cold rolling is a good finishing treatment in the production of plate and sheet. The sheets of steel for car bodies are finished by cold rolling because a good surface finish is essential in this product.

Metal Forging

Forging is typified by countless generations of blacksmiths with their hammers and anvils. Besides still being used for special 'hand-made' items, this type of forging is similar to that used, on a somewhat larger scale, for the initial rough shaping of hot metal ingots. Forging is particularly good at making 3D solid shapes. The basic types of forging processes are shown in Figure.

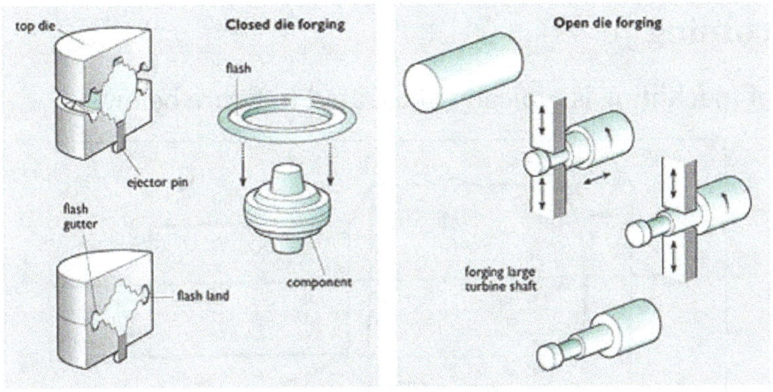

Forging processes

For example, the first stage in making a large roller or shaft would be to forge a large billet as sketched in Figure. The process consists of a large succession of bites between a pair of dies in an hydraulic press, with the ingot moved between each bite. As the ingot is moved through the dies it is reduced to a more manageable size before final shaping – this process is known as open die forging.

In closed die forging components are made in one action, being squeezed between upper and lower shaped dies as shown in Figure. There is usually a small amount of excess material which is forced out of the die cavity as flash. This must then be removed from the component. The force needed to close the dies together is dependent both on the size of the component and the temperature, since as we noted earlier, the flow stress reduces as the forging temperature increases. The quality of the surface finish of the forging decreases with temperature, however, because of increased surface oxidation.

Machining

Machining is a manufacturing term encompassing a broad range of technologies and techniques. It can be roughly defined as the process of removing material from a workpiece using power-driven machine tools to shape it into an intended design. Most metal components and parts require some form of machining during the manufacturing process. Other materials, such as plastics, rubbers, and paper goods, are also commonly fabricated through machining processes.

Purpose of Machining

Most of the engineering components such as gears, bearings, clutches, tools, screws and nuts etc. need dimensional and form accuracy and good surface finish for serving their purposes. Preforming like casting, forging etc. generally cannot provide the desired accuracy and finish. For that such preformed parts, called blanks, need semi-finishing and finishing and it is done by machining and grinding. Grinding is also basically a machining process.

Machining to high accuracy and finish essentially enables a product:

- Fulfill its functional requirements.
- Improve its performance.
- Prolong its service.

Principle of Machining

The basic principle of machining is typically illustrated in figure below.

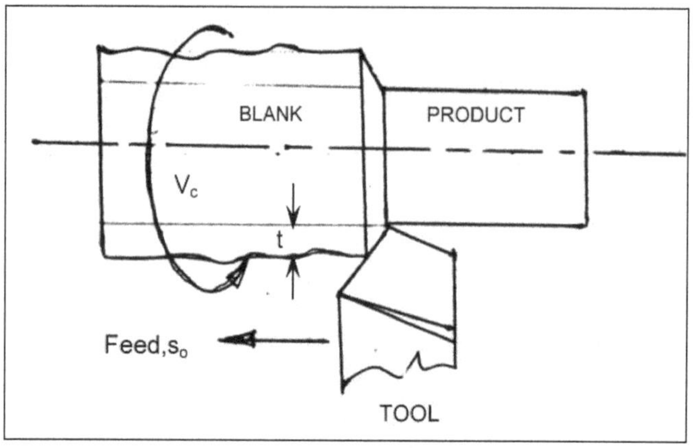

Principle of machining (turning).

A metal rod of irregular shape, size and surface is converted into a finished rod of desired dimension and surface by machining by proper relative motions of the tool-work pair.

Definition of Machining

Machining is an essential process of finishing by which jobs are produced to the desired dimensions and surface finish by gradually removing the excess material from the preformed blank in the form of chips with the help of cutting tools moved past the work surfaces.

Machining Requirements

The essential basic requirements for machining work are schematically illustrated in figure below.

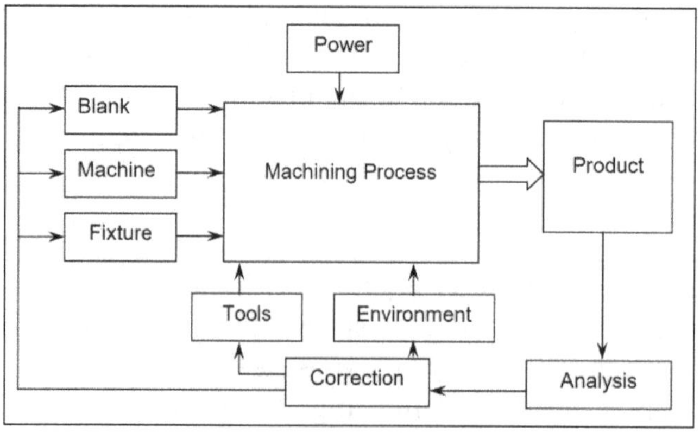

Requirements for machining.

The blank and the cutting tool are properly mounted (in fixtures) and moved in a powerful device called machine tool enabling gradual removal of layer of material from the work surface resulting in its desired dimensions and surface finish. Additionally some environment called cutting fluid is generally used to ease machining by cooling and lubrication.

Basic Functions of Machine Tools

Machine Tools basically produce geometrical surfaces like flat, cylindrical or any contour on the preformed blanks by machining work with the help of cutting tools. The physical functions of a Machine Tool in machining are:

- Firmly holding the blank and the tool
- Transmit motions to the tool and the blank
- Provide power to the tool-work pair for the machining action.
- Control of the machining parameters, i.e., speed, feed and depth of cut.

Types of Machining Tools

There are many types of machining tools, and they may be used alone or in conjunction with other tools at various steps of the manufacturing process to achieve the intended part geometry. The major categories of machining tools are:

- Boring tools: These are typically used as finishing equipment to enlarge holes previously cut into the material.
- Cutting tools: Devices such as saws and shears are typical examples of cutting implements. They are often used to cut material with predetermined dimensions, such as sheet metal, into a desired shape.
- Drilling tools: This category consists of two-edged rotating devices that create round holes parallel to the axis of rotation.
- Grinding tools: These instruments apply a rotating wheel to achieve a fine finish or to make light cuts on a workpiece.
- Milling tools: A milling tool employs a rotating cutting surface with several blades to create non-circular holes or cut unique designs out of the material.
- Turning tools: These tools rotate a workpiece on its axis while a cutting tool shapes it to form. Lathes are the most common type of turning equipment.

Types of Burning Machining Technologies

Welding and burning machine tools use heat to shape a workpiece. The most common types of welding and burning machining technologies include:

- Laser cutting: A laser machine emits a narrow, high-energy beam of light that effectively melts, vaporizes, or burns material. CO_2 and Nd:YAG lasers are the most common types used in machining. The laser cutting process is well-suited for shaping steel or etching patterns into a piece of material. Its benefits include high-quality surface finishes and extreme cutting precision.

- Oxy-fuel cutting: Also known as gas cutting, this machining method employs a mixture of fuel gases and oxygen to melt and cut away material. Acetylene, gasoline, hydrogen, and propane frequently serve as gas media due to their high flammability. This method's benefits include high portability, a low dependence on primary power sources, and the ability to cut thick or hard materials, such as sturdy steel grades.
- Plasma cutting: Plasma torches fire an electrical arc to transform inert gas into plasma. This plasma reaches extremely elevated temperatures and is applied to the workpiece at high speed to melt away unwanted material. The process is often used on electrically conductive metals that require a precise cut width and minimal prep time.

Types of Erosion Machining Technologies

While burning tools apply heat to melt excess stock, erosion machining devices use water or electricity to erode material off the workpiece. The two main types of erosion machining technologies are:

- Water jet cutting: This process uses a high-pressurized stream of water to cut through material. Abrasive powder may be added to the water stream to facilitate erosion. Water jet cutting is typically used on materials that can suffer damage or deformation from a heat affected zone.
- Electric discharge machining (EDM): Also known as spark machining, this process uses electric arcing discharges to create micro-craters that rapidly result in complete cuts. EDM is used in applications requiring complex geometrical shapes in hard materials and at close tolerances. EDM requires the base material to be electrically conductive, which limits its use to ferrous alloys.

CNC Machining

Computer numerical control, or CNC, machining is a computer-aided technique that can be used in conjunction with a broad range of equipment. It requires software and programming, usually in the G-code language, to guide a machining tool in shaping the workpiece according to preset parameters. As opposed to manually guided methods, CNC machining is an automated process. Some of its benefits include:

- High production cycles: Once the CNC machine has been properly coded, it usually needs minimal maintenance or downtime, allowing for a faster production rate.
- Low manufacturing costs: Due to its turnover speed and low manual labor requirements, CNC machining can be a cost-efficient process, particularly for high-volume production runs.
- Uniform production: CNC machining is typically precise and yields a high level of design consistency among its products.

Precision Machining

Any machining process that requires unusually small cutting tolerances (between 0.013 mm and 0.0005 mm, as a rule of thumb) or surface finishes finer than 32T may be considered a form of precision machining. Like CNC machining, precision machining can be applied to a wide number of fabrication methods and tools. Factors such as stiffness, damping, and geometric accuracy can influence the exactness of a precision tool's cut. Motion control and the machine's ability to respond at rapid feed rates are also important in precision machining applications.

Joining

The joining of materials is an important technology in many manufacturing industries. Most products, machines or structures are assembled and fastened from parts, and the joining of these parts may be achieved through rivets, seaming, clamping, soldering, brazing, welding and the use of adhesives. Every joining approach has particular design needs, while certain joint needs may propose a particular joining approach. Design for assembly, and fastener selection apply their own specifications.

Bolting is a standard fastening method, for instance, but welding may cut down the weight of assemblies. Naturally, joints intended for the two approaches would differ tremendously.

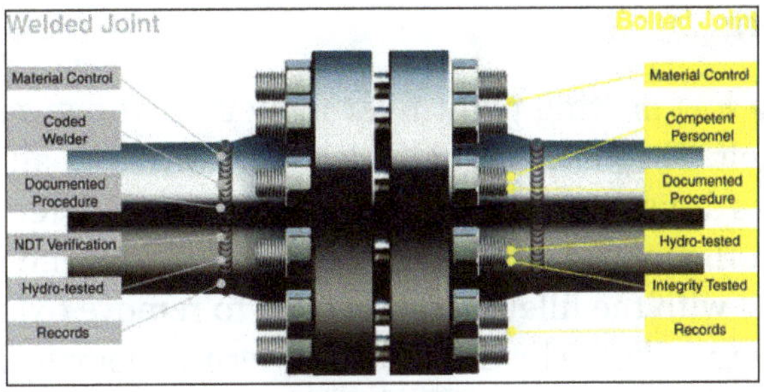

However, all joint patterns must consider features such as load factors, assembly effectiveness, operating surroundings, overhaul and upkeep, and the materials chosen.

Welding is generally a cost-effective approach to fabricate. It doesn't require overlapping materials, and so it removes excess weight brought on by other fastening methods. Fasteners don't have to be purchased and stored in stock. Welding also can minimize costs related to extra parts, for example angles mounted between parts.

Adhesive Bonding

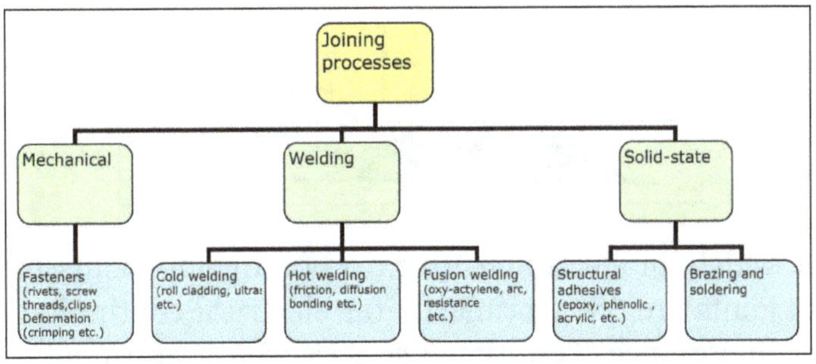

Adhesive bonding is applicable to all classes of materials and is increasingly used for metals. Extensively used to join polymer (both thermoplastic and thermoset) and composite components. Several classes of polymer are used as adhesives: e.g. epoxies, phenolics, acrylics, polyurethanes. Adhesives are mostly thermosets, made by mixing two components (resin and crosslinking initiator) before use. Some require curing at modest temperatures (100 to 200 °C).

Adhesive joints require careful surface cleaning and preparation, and selection of adhesives which are compatible with the component materials. Adhesives are often susceptible to take-up of moisture, which can affect joint integrity and can lead to degradation of properties in service. Joint designs are usually some form of lap joint, because the adhesives are much stronger in shear than in tension (which would produce hydrostatic tension leading to void growth and failure).

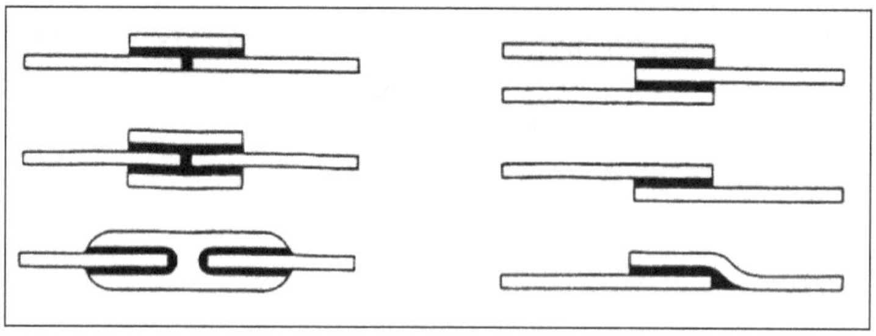

Brazing and Soldering

In brazing and soldering a low melting point filler alloy is used between the parts being joined, which themselves never melt. The joint is usually designed with a small clearance so that the molten solder or brazing alloy is drawn into the gap by capillary (surface tension) forces. It is therefore particularly important that the molten filler wets the two surfaces, and a flux (which is an aggressive chemical agent applied with the filler metal) is used to remove oxide from the metal surfaces and allow good wetting by the filler. The difference in terminology is essentially associated with melting temperature: so-called 'soft solders' melt below 450 °C, and brazing alloys or 'hard solders' melt above 450 °C.

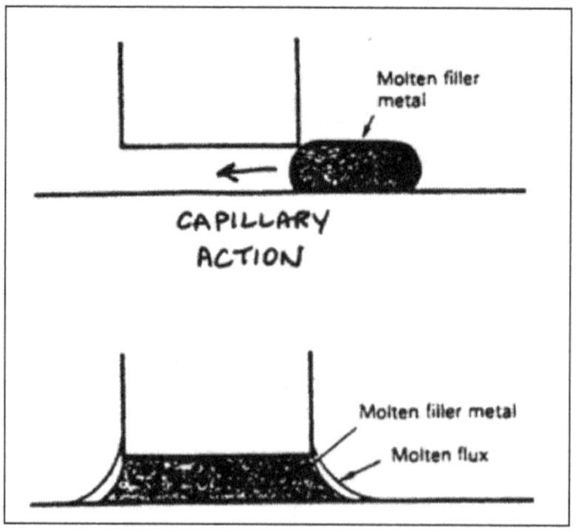

Soldering is used very widely in electronics (e.g. to bond components to circuit boards, wires to contacts, etc.) these joints are often primarily for electrical continuity, with mechanical performance of secondary importance. Traditional solders were lead-tin alloys (e.g. 37%Pb-63%Sn eutectic melts at about 180 °C) but for environmental reasons these have largely been replaced by lead-free solders based on tin-copper-silver alloys, which are more expensive (e.g. the eutectic Sn-3.5%Ag-0.9%Cu melts at 217 °C).

Brazing is used for mechanical purposes and often uses a copper-zinc or copper-silver alloy. With appropriate filler alloys and fluxes (and in some cases an inert atmosphere or vacuum environment) brazing methods can join a wide range of metals, including steels, stainless steels, aluminium alloys as well as ceramics.

Both brazing and soldering suffer from the disadvantages that the strength of the joint is limited by the strength of the filler alloy, and also that this is often much more expensive than the metal of the parts being joined. However, for many applications, and for joining dissimilar metals, or metals to ceramics, the processes are very valuable and since the temperature involved is lower than for welding they usually produce less distortion.

Welding Processes for Metals

Welding processes join two components either without any filler material or with filler material of very similar composition (and melting point) to the materials being joined. The strength of a welded joint can approach that of the neighbouring material. Contamination present on the original surfaces (e.g. oxide) is broken up or removed by melting or in some processes by plastic deformation.

1. COLD welding ($T < 0.3\ T_m$): uses low temperature plastic deformation to provide good metal-metal contact and disrupt oxide films.

Examples:

- Roll cladding: To clad Al alloys with pure Al for corrosion resistance.
- Ultrasonic welding: For thermoplastic polymers (in which the T rise is significant in promoting interdiffusion and bonding), and for making metallic bonds in light components e.g. Al or Au wire contacts between i.e. chips and mounting substrates, other electrical contacts.

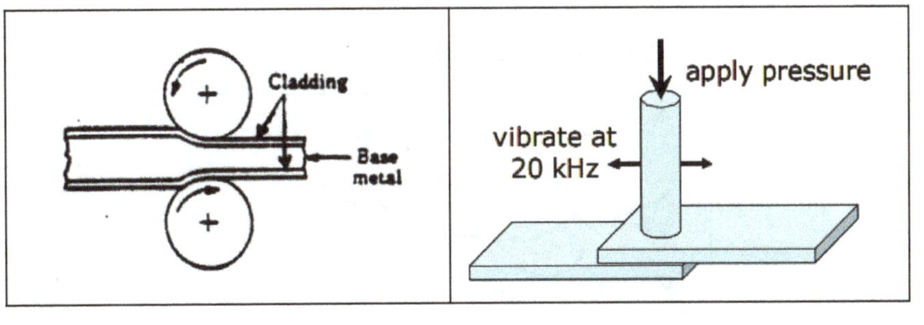

2. HOT welding ($0.5\ T_m < T < T_m$): uses hot plasticity/creep/diffusion to provide good metal-metal contact.

Examples:

- Friction welding: rotating parts are forced together producing simultaneous frictional heat generation and local plastic flow. Mainly used for joints with circular symmetry (pipes, shafts etc) – e.g. Automotive components (steering column, drive shafts).

- Diffusion bonding: slower process in which heat and pressure lead to interdiffusion across the joint interface. Used for Ni-based superalloys and Ti alloys in aerospace (also for ceramics). Process analogous to sintering of powders.

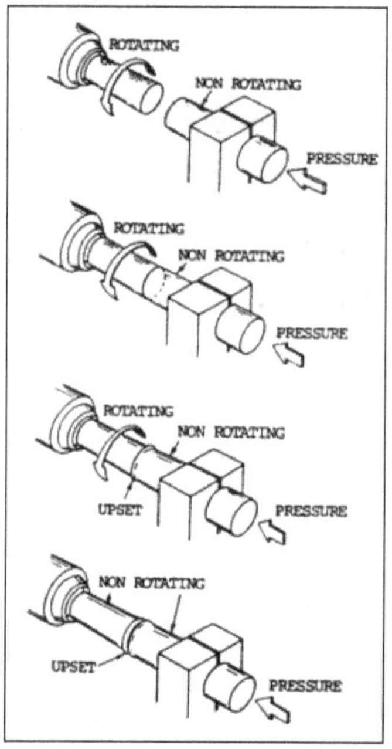

3. FUSION welding ($T > T_m$): components to be joined are melted along their common interface by an external heat source. Filler metal may be used to fill the gap.

The main fusion welding processes can be classified in terms of the source of heat as:

- Resistance welding
- Gas welding
- Arc welding
- Power beam welding (laser or electron beam).

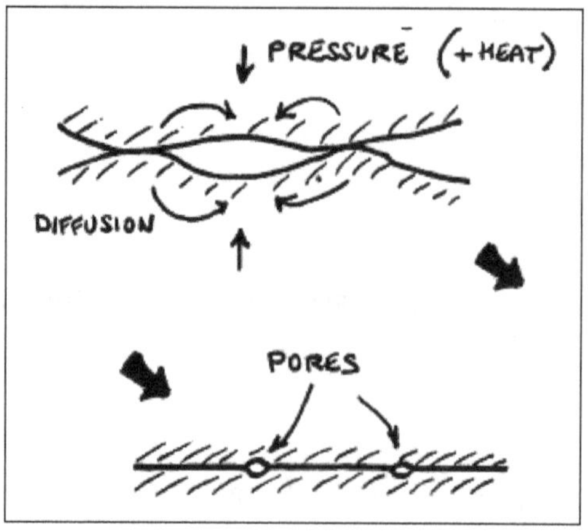

Important factors in selecting a welding method are the material to be joined, its geometry (e.g. plate, tubular etc), the quality and strength requirements, degree of mechanization appropriate, portability required, and capital cost of equipment.

Resistance Welding

Uses electric current to heat the metal locally by power dissipation in resistance at contact region. Often used for sheet material. Two sheets are pressed together in a lap joint between two electrodes. A very quick and effective method for producing 'spot' or 'seam' welds (by using rollers on each side of the joint).

Easily automated, and commonly used in robotic car body/substructure fabrication.

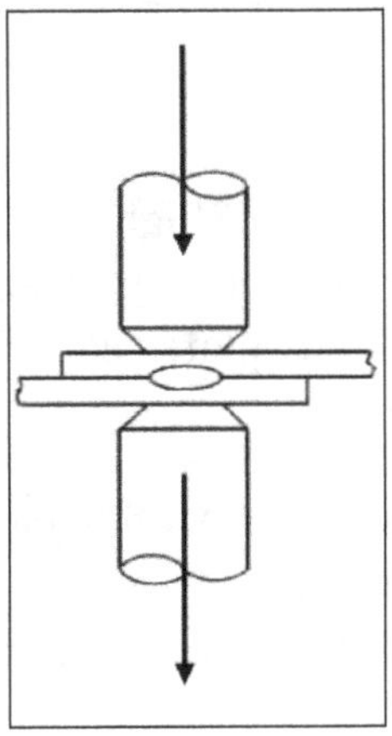

Gas Welding

(Oxy-acetylene welding) A simple and cheap method (in terms of equipment) which usually uses a manually-operated gas torch burning acetylene (= ethyne: C_2H_2) in oxygen. The filler rod is supplied manually to the melt pool and the molten metal is prevented from oxidizing by the combustion gases (which can be adjusted to produce a reducing i.e. non-oxidizing flame).

Arc Welding

An electrical arc discharge is generated between electrode and workpiece using high voltage/current. A meltable inorganic material (flux) or inert shrouding gas (e.g. argon) is provided to prevent oxidation of the melt pool. The processes differ in the source of the filler material (which may be provided as a consumable rod or wire electrode, or supplied separately into a melt pool with a non-consumable electrode used to form the arc). Processes can be manual, semi-automated or robotic. Arc welding processes are very common in structural uses of steel and Al.

Common types of arc welding process include:

- Manual metal arc (MMA) with a flux-covered consumable electrode and no shielding gas (i.e. 'self-shielded'). The flux melts and floats on the weld metal, forming a protective layer of slag which is later removed.

- Metal-inert-gas (MIG) which uses a consumable electrode (rod or wire), to provides the filler material, and a supply of inert shielding gas.
- Tungsten-inert-gas (TIG) which uses a non-consumable tungsten electrode, inert gas shielding and a separate filler rod.

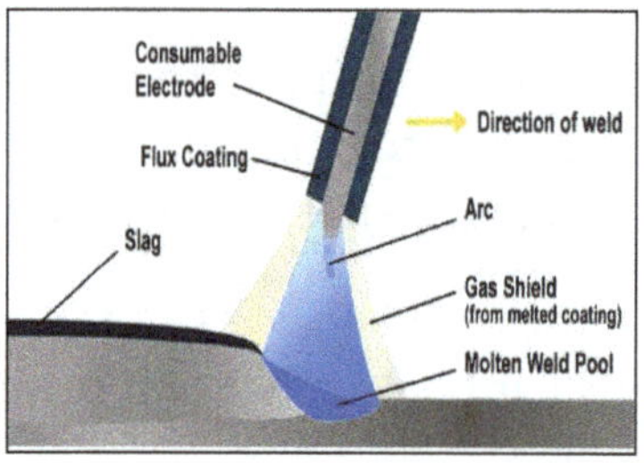

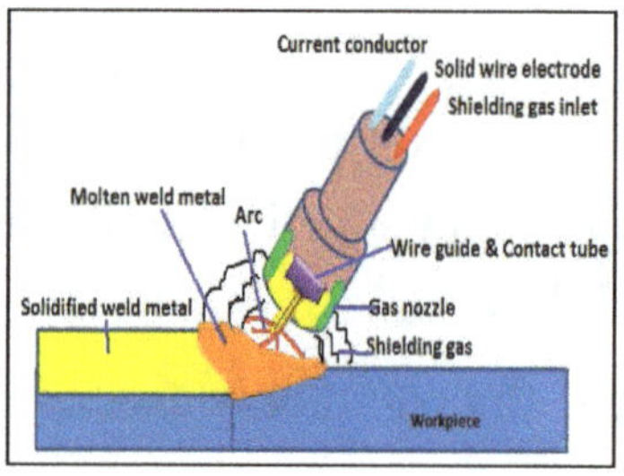

Power Beam Processes

Laser and electron beam welding use a beam of high power density as the heat source. The beam vapourises a small surface spot, giving high energy absorption from the beam - a deep narrow vapour cavity forms called a 'keyhole'. Thick plates can be welded without filler in a single pass, giving narrow melt and heat-affected zones.

Electron beam welding is especially good for very thick plate but requires vacuum or partial vacuum; laser welding only requires a shrouding gas.

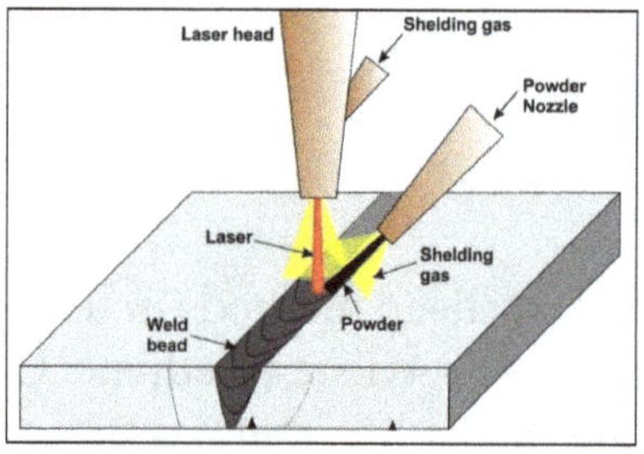

Lasers can also be used at a lower power density for welding sheet materials - here the beam only melts the material and produces a melt pool similar to an arc process (i.e. no keyhole is formed).

Example: Car bodywork, including 'tailored blanks' - panels of different steels joined into a single flat sheet prior to forming by pressing into curved body panels. Readily automated, ideal for robotic control.

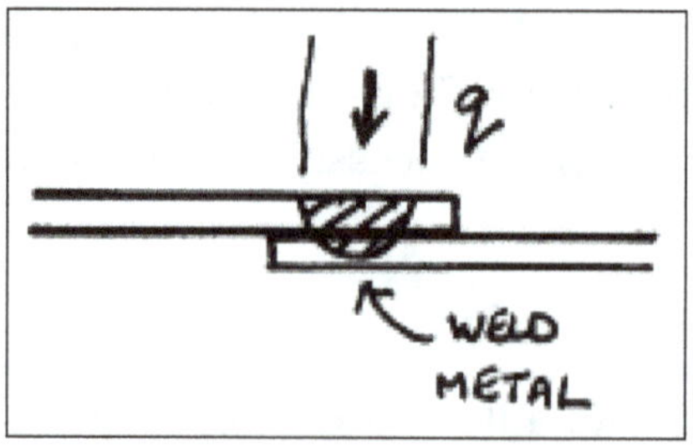

A significant advantage of laser welding over other processes is the ease with which the energy can be supplied, concentrated and switched on and off at exactly the region where is needed: the steerability of the beam and fine control of power allow precise delivery of heat input, giving low distortion and excellent reproducibility and quality control of the process.

Metallurgical Effects in Welding

The 'weld metal' is formed by solidification of the molten pool in fusion welding. It is like a miniature casting, with new metal being melted in from the filler and the leading edge of the pool, and solidification occurring around the trailing edge of the pool. The shape of the pool depends on the heat input and the welding speed - which control the shape of the temperature distribution around the weld (including the melt isotherm).

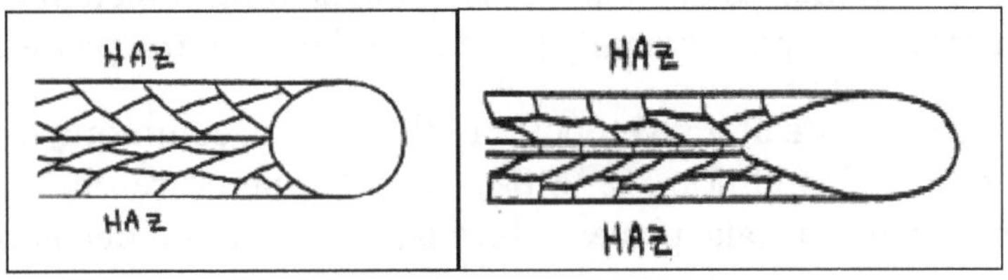

(HAZ = heat affected zone)

Epitaxial solidification occurs as grains grow into the melt pool from the unmelted regions.

Segregation can occur, with solute and impurities concentrated on the centre-line. Segregation of sulphur for example gives brittle grain boundaries– solidification cracking. Hydrogen cracking can also occur if a source of hydrogen is present (e.g. from damp electrodes in MMA welding).

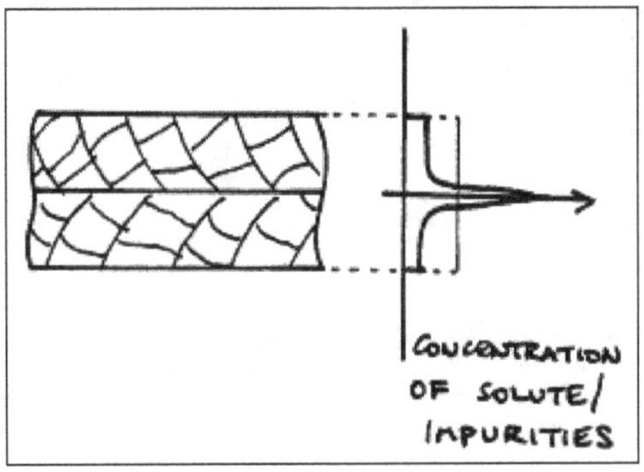

The heat-affected zone (HAZ) which surrounds the weld metal (but does not actually melt) plays a very important role in determining the properties and performance of a fusion weld. Several processes can occur in the HAZ during the heating and cooling cycle:

- Recovery, recrystallization and grain growth.
- Precipitate coarsening/dissolution.
- Phase transformations.

The HAZ microstructure depends on:

- The alloy composition.
- Its initial microstructure.

- The welding process (which determines the heat input, size of the melted zone, and the heating/cooling cycle).

The HAZ in Steels

The most important property in fusion welding of steels is the final toughness. Since there is an inverse correlation between hardness and toughness in steels, it is usual to specify a maximum allowable final hardness in the HAZ.

CCT diagrams assume that we start with 100% austenite and an initial temperature around 800-850 °C. In a weld, transformation to austenite during the heating transient may not be complete. But the CCT diagram provides a good starting point to understand the process.

In steel welding, the cooling rate at a given point in the HAZ is commonly described by the time taken to cool between 800 °C and 500 °C: '$\Delta t8\text{-}5$'. This temperature interval is where the most important phase transformations take place: if the austenite does not decompose then martensite is likely to form, leading to brittleness.

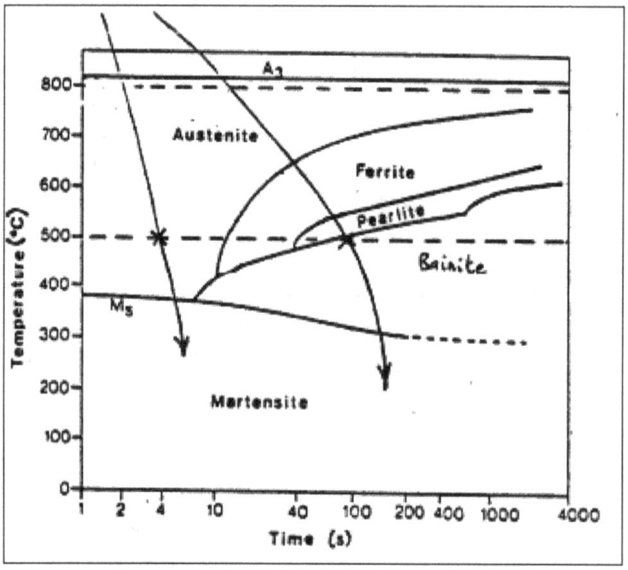

A high cooling rate is equivalent to a small $\Delta t_{8\text{-}5}$.

- Small $\Delta t_{8\text{-}5}$: fine structure, more likely to form martensite.
- Large $\Delta t_{8\text{-}5}$: coarse structure, no martensite, not brittle.

$\Delta t_{8\text{-}5}$ can be related to the welding parameters (heat input, plate thickness etc) and position in the HAZ.

Typical values:

Welding process	Δt8-5	dT/dt (from 800 to 500 °C)
Metal arc	100 s	3 K s^{-1}
EB/laser	0.1 – 1 s	300-3000 K s^{-1}

The hardness in the HAZ and weld metal depends on:

- The amount of austenite formed on heating.
- The products of the austenite decomposition (i.e. Proportions of martensite, ferrite/pearlite, bainite).
- Changes to the original grains which do not change to austenite.

Carbon equivalent, CE, is often used as an indicator of weldability.

For strong, tough welds in steels:

- Use low CE steels (< 0.4).
- Use a fine-grained 'microalloyed' steel (to prevent grain growth).
- Preheat to ~ 200 °C to reduce subsequent cooling rates, and reduce the likelihood of martensite formation.
- Post-weld heat treat at 650 °C - this tempers any martensite (and also relieves residual stresses).

HAZ in Aluminium Alloys

Many heat-treatable and non-heat-treatable Al alloys are weldable, but suffer some strength loss in the HAZ. The thermal cycle may be regarded as a brief annealing treatment the extent of microstructural change depends on the softening mechanisms in the two classes of alloy.

A weld in a non-heat-treatable wrought alloy, which will produce a cast microstructure, will be inevitably weaker than the surrounding material, as is the HAZ (which has been annealed by the heating cycle).

For an age-hardening (heat-treatable) aluminium alloy (e.g. Al-Mg-Si (6000 series) or Al-Mg-Zn-Cu (7000 series)) it is possible to recover much of the strength after welding. Consider a weld in a 6000 or 7000 series alloy, peak aged to T6 temper. Immediately after welding the hardness profile is as shown on the left.

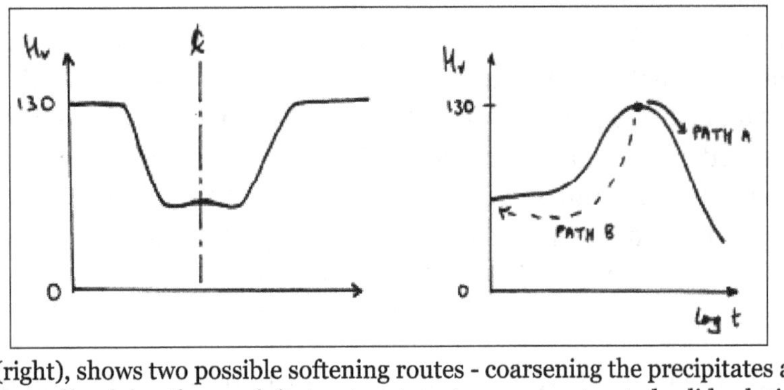

The ageing curve (right), shows two possible softening routes - coarsening the precipitates and overaging the alloy (path A), or dissolving the precipitates, to return to a supersaturated solid solution (path B).

In most weldable aluminium alloys, we get re-solution (i.e. path B), a process known as reversion. This is very important, since it offers the potential to restore strength to the HAZ.

Welds cannot be aged at elevated temperatures to give the full T6 strength as that would cause the surrounding material to over-age and soften. They can however be naturally aged at room temperature - the 7000 series alloys in particular age-harden rapidly at room temperature to a substantial proportion of their original unwelded strength, as sketched below.

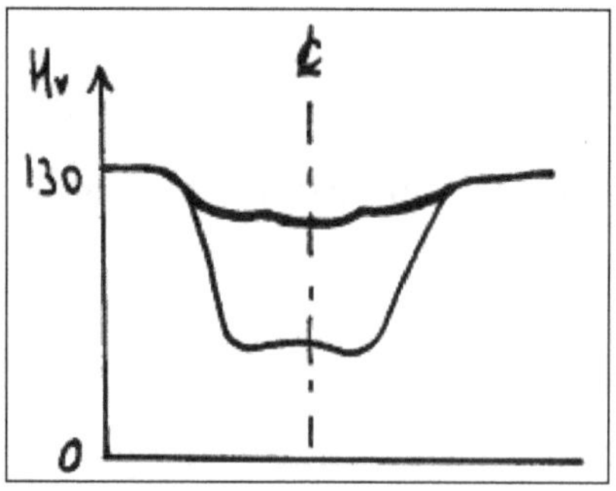

Design of Shaft

Shaft is a common machine element which is used to transmit rotary motion or torque. It generally has circular cross-section and can be solid or hollow. Shafts are supported on the bearings and transmit torque with the help of gears, belts and pulleys etc. Shafts are generally subjected to bending moment, torsion and axial force or a combination of these three. So the shafts are designed depending upon the combination of loads it is subjected to. Spindle stub and axle are some important types of shaft. Small shaft is called spindle. Shaft integral part of the prime mover is called stub shaft. An axle is a non-rotating member that carries no torque and is used to support rotating wheels, pulleys etc. And therefore is subjected to bending moment only.

Shaft Materials

Hot-rolled plain carbon steel is the least expensive material used for shafts. These essentially require machining to remove the scales of hot rolling process. Cold rolled plain carbon steel provides better yield strength and endurance strength but the cold working induces residual stresses. Surface is smooth in this case and amount of machining therefore is minimal. It is used for general purpose transmission shafts. When a shaft is to work under severe loading and corrosive conditions and require more strength, alloy steels are used, generally having Ni, Cr, Mo and V as alloying elements. Alloy steels are expensive. Sometimes shafts are heat treated to improve hardness and shock resistance and surface hardening techniques are also used if high wear resistance is the requirement. As the shafts transmitting power are subjected to fatigue loading, therefore higher factor of safety of 3 to 4 is used on the basis of yield strength for static load analysis.

Shaft Designing

Shafts are designed on the basis of strength or rigidity or both. Design based on strength is to ensure that stress at any location of the shaft does not exceed the material yield stress. Design based on rigidity is to ensure that maximum deflection (because of bending) and maximum twist (due to torsion) of the shaft is within the allowable limits. Rigidity consideration is also very important in some cases for example position of a gear mounted on the shaft will change if the shaft gets deflected and if this value is more than some allowable limit, it may lead to high dynamic loads and noise in the gears.

In designing shafts on the basis of strength, the following cases may be considered:

- Shafts subjected to torque.
- Shafts subjected to bending moment.
- Shafts subjected to combination of torque and bending moment.
- Shafts subjected to axial loads in addition to combination of torque and bending moment.

Shafts Subjected to Torque

Maximum shear stress developed in a shaft subjected to torque is given by,

$$\tau = \frac{Tr}{J} \leq [\tau]$$

where T = Twisting moment (or torque) acting upon the shaft,

J = Polar moment of inertia of the shaft about the axis of rotation

$= \frac{\pi d^4}{32}$ for solid shafts with diameter d

$= \frac{\pi \left(d_o^4 - d_o^4\right)}{32}$ for hollow shafts with do and di as outer and inner diameter.

r = Distance from neutral axis to the outer most fibre = d/2 (or d_o/2).

So dimensions of the shaft subjected to torque can be determined from above relation for a known value of allowable shear stress, [τ].

Shafts Subjected to Bending Moment

Maximum bending stress developed in a shaft is given by,

$$\sigma_b = \frac{M\ y}{I} \leq [\sigma_t]$$

where M = Bending Moment acting upon the shaft, I = Moment of inertia of cross-sectional area of the shaft about the axis of rotation $= \frac{\pi d^4}{64}$ for solid shafts with diameter d$= \frac{\pi \left(d_o^4 - d_o^4\right)}{64}$ for hollow shafts with d_o and d_i as outer and inner diameter.

Y = Distance from neutral axis to the outer most fibre = d / 2 (or d_o/2).

So dimensions of the shaft subjected to bending moment can be determined from above relation for a known value of allowable tensile stress.

Shafts Subjected to Combination of Torque and Bending Moment

When the shaft is subjected to combination of torque and bending moment, principal stresses are calculated and then different theories of failure are used. Bending stress and torsional shear stress can be calculated using the above relations.

$$\tau = \frac{T\,r}{J} = \frac{T\dfrac{d}{2}}{\dfrac{\pi}{32}d^4} = \frac{16T}{\pi d^3}$$

$$\sigma_b = \frac{M\,y}{I} = \frac{M\dfrac{d}{2}}{\dfrac{\pi}{64}d^4} = \frac{32M}{\pi d^3}$$

Maximum Shear Stress Theory

Maximum shear stress is given by,

$$\tau_{\max} = \sqrt{\left(\frac{\sigma_b}{2}\right)^2 + (\tau)^2} = \sqrt{\left(\frac{16M}{\pi d^3}\right)^2 + \left(\frac{16\,T}{\pi d^3}\right)^2} = \frac{16}{\pi d^3}\sqrt{M^2+T^2} \le [\tau]$$

$\sqrt{M^2+T^2}$ is called equivalent torque, T_e, such that:

$$\tau_{\max} = \frac{T_e\,r}{J} \le [\tau].$$

Maximum Principal Stress Theory

Maximum principal stress is given by,

$$\sigma = \frac{\sigma_b}{2} + \sqrt{\left(\frac{\sigma_b}{2}\right)^2 + (\tau)^2} = \frac{16\,M}{\pi d^3} + \sqrt{\left(\frac{16\,M}{\pi d^3}\right)^2 + \left(\frac{16\,T}{\pi\,d^3}\right)^2} = \frac{16}{\pi d^3}\left[M + \sqrt{M^2+T^2}\right] \le [\sigma_t]$$

$\left[M + \sqrt{M^2+T^2}\right]$ is called equivalent bending moment, M_e, such that

$$\sigma = \frac{M_e\,y}{I} \le [\sigma_t].$$

A.S.M.E. Code for Shaft Design

According to A.S.M.E. code, the bending and twisting moment are to be multiplied by factors k_b

and k_t respectively, to account for shock and fatigue in operating condition. Therefore, if the shaft is subjected to dynamic loading, equivalent torque and equivalent bending moment will become:

$$T_e = \sqrt{k_b M^2 + k_t T^2} \quad \text{and} \quad M_e = \left[k_b M + \sqrt{k_b M^2 + k_t T^2} \right]$$

Table: Values of k_b and k_t for different types of loading.

	k_b	k_t
Gradually applied load	1.5	1.0
Suddenly applied load (minor shock)	1.5-2.0	1.0-1.5
Suddenly applied load	2.0-3.0	1 .5-3.0

Shafts Subjected to Axial Loads in addition to Combination of Torque and Bending Moment:

Tensile Stress due to axial load is given by,

$$\sigma_t = \frac{P}{A}$$

where, P = axial load acting on the shaft,

A = cross-sectional area of the shaft.

As nature of the bending stress and this axial stress is same, these can be vectorially added for any location on the shaft, so as to get the resultant tensile/compressive stress, which can then be used to find the principal stresses in the shaft.

Design of Shaft on the basis of Rigidity

Torsional Rigidity

For a shaft subjected twisting moment, the angle of twist is given by,

$$\theta = \frac{TL}{GJ} \leq [\theta]$$

Where, T = Torqe applied

L = Length of the shaft.

J = Polar moment of inertia of the shaft about the axis of rotation.

G = Modulus of rigidity of the shaft material.

Therefore for the known values of T, L and G and allowable value of angle of twist, diameter of the shaft can be calculated.

Lateral Rigidity

Bending moment acting on any shaft is given by,

$$M = EI\frac{d^2y}{dx^2}$$

Integrating this equation twice with respect to x and applying the boundary conditions, y can be calculated. y should be ≤ allowable value of deflection, [y].

A.S.M.E. Code for Shaft Design

According to A.S.M.E. code, the bending and twisting moment are to be multiplied by factors k_b and k_t respectively, to account for shock and fatigue in operating condition. Therefore, if the shaft is subjected to dynamic loading, equivalent torque and equivalent bending moment will become:

$$T_e = \sqrt{k_b M^2 + k_t T^2} \qquad and \quad M_e = \left[k_b M + \sqrt{k_b M^2 + k_t T^2}\right]$$

Table: Values of k_b and k_t for different types of loading.

	k_b	k_t
Gradually applied load	1.5	1.0
Suddenly applied load (minor shock)	1.5-2.0	1.0-1.5
Suddenly applied load	2.0-3.0	1.5-3.0

Clutches

A clutch is a machine member used to connect a driving shaft to a driven shaft so that the driven shaft may be started or stopped at will, without stopping the driving shaft. A clutch is a mechanical device that provides for the transmission of power (and therefore usually motion) from one component (the driving member) to another (the driven member) when engaged, but can be disengaged.

Friction Clutch

The friction clutch is used to transmit power of shafts and machines which must be started and stopped frequently.

Friction surfaces of a clutch remain in contact to each other by applying an axial thrust or load w.

where, r_2 = External radius of friction faces,

m = Coefcient of friction,

p = Intensity of axial pressure.

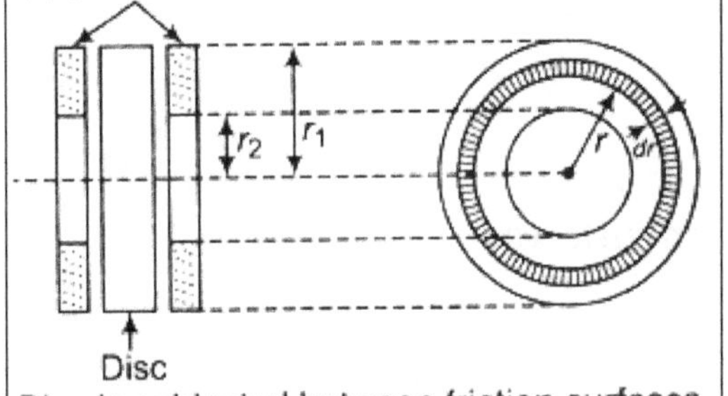

Disc is subjected between friction surfaces

Considering Uniform Pressure

Let us consider r_1 and r_2 are the external and internal radius of friction faces respectively. The axial thrust w with which the friction surfaces held together, then the uniform pressure p can be evaluated as.

$$p = \frac{w}{\pi\left(r_1^2 - r_2^2\right)}$$

Total frictional torque given in this case,

$$T = \frac{2}{3}\mu w\left(\frac{r_1^3 - r_2^3}{r_1^2 - r_2^2}\right) = mw\ R_m$$

where, Rm = Mean radius of friction surfaces

$$R_m = \frac{2}{3}\left(\frac{r_1^3 - r_2^3}{r_1^2 - r_2^2}\right).$$

Considering Uniform Wear

Assume, p is the pressure on friction surfaces at a distance r from the axis of the clutch.

According uniform wear theory,

pr = constant $= c \Rightarrow p = \dfrac{w}{2\pi\ r\left(r_1 - r_2\right)}$

Total frictional torque acting on clutch,

$$T = \frac{1}{2}\mu w\left(r_1 + r_2\right) = \mu w R_m$$

where R_m = Mean radius of friction surfaces $= \dfrac{r_1 - r_2}{2}$

In uniform wear theory, Maximum pressure acts at the inner radius and minimum pressure acts at the outer radius.

$$P_{max} \times r_2 = c_2 P_{min} \times r_1 = c$$

$$\frac{P_{max}}{P_{max}} = \frac{r_1}{r_2}$$

Average pressure on the friction surfaces,

$$P_{av} = \frac{w}{\pi\left(r_1^2 - r_2^2\right)}.$$

Multiple Disc Clutch

In multiple disc clutch, the inside discs are fastened to the driven shaft to permit axial motion. The outside discs are held by bolts and are fastened to the housing which is keyed to the driving shaft.

Number of pairs of contact surfaces,

$$n = n_1 + n_2 - 1$$

where, n_1 = Number of discs on the driving shaft,

n_2 = Number of discs on the drive shaft.

Total frictional torque acting on the frictional surface,

where, $R_m = \frac{2}{3}\left(\frac{r_1^2 - r_1^2}{r_1^2 - r_2^2}\right)$ [in case of uniform pressure],

$= \frac{1}{2}(r_1 - r_2)$ [in case of uniform wear]

where, r_1 and r_2 are outer and inner radii of the friction plates.

Cone Clutch

In cone clutch, driver is keyed to the driving shaft by a sunk key and has an inside conical surface or face which exactly ts into the outside conical surface of the driven.

Total torque on the clutch,

$T = \frac{2}{3}\mu w \cos ec\left[\frac{r_1^3 - r_2^3}{r_1^2 - r_1^2}\right]$ (for uniform pressure)

$= \frac{1}{2}\mu w \cos ec\ \alpha\left(r_1 + r_2\right)$ (for uniform wear)

α = Semi angle of cone or face angle of the cone,

$$w_n = \frac{w}{\sin\alpha}$$

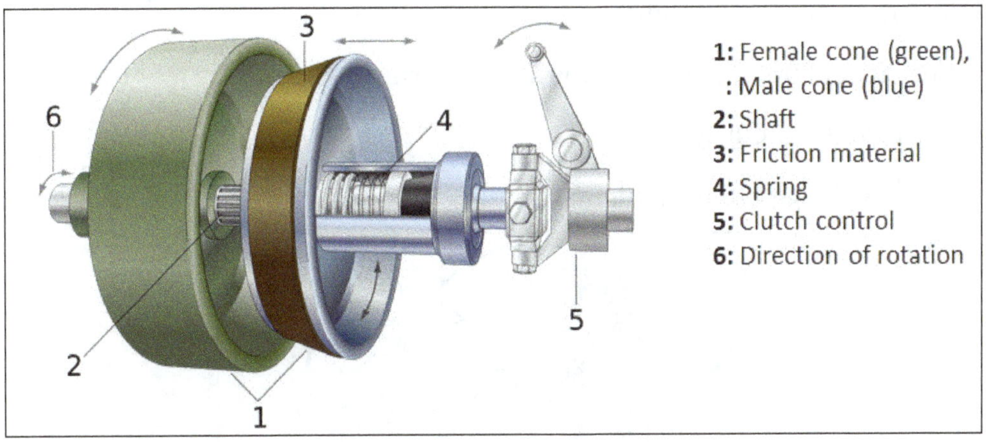

where, w = Axial load or thrust

- Axial force required for engaging the clutch,

$$= w_n \left(\sin\alpha + \mu \cos\alpha\right)$$

- Axial force required to disengaged the clutch

$$w_d = w_n \left(\mu \cos\alpha - \sin\alpha\right)$$

- If face width b and mean radius of cone clutch is R_m.

Then, $R_m = \left(\frac{r_1 - r_2}{2}\right)$

Then, $R_n = \frac{W}{2\pi R_m b \sin\alpha}, T = 2\pi\mu.P_n R_m b$

- Power transmitted by clutch $= \frac{2\pi NT}{60}$

Where, N = Speed of shaft (rpm)

T = Frictional torque.

Centrifugal Clutch

It consists of number of shoes on the inside of a rim of the pulley. The outer surface of the shoes is covered with a friction material. These shoes, which can move radially in guides, are held against the boss on the driving shaft by means of spring. The springs exert a radially inward force which is assumed constant.

Let,

M = Mass of each shoe.

r = Distance between centre of gravity of the shoe and the centre of spider.

N = Running speed of the pulley.

m = Coefcient of friction between the shoe and rim.

R = Inner radius of pulley rim.

n = Number of shoes in the pulley rim.

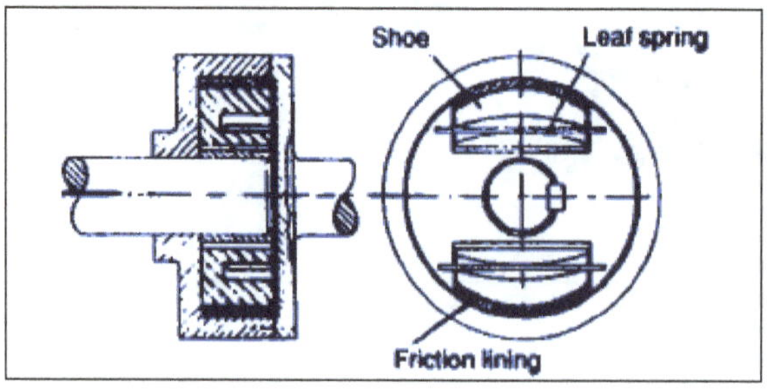

When engagement starts then, the spring force equal to the centrifugal force acts at shoe at that time.

$$Ps = m\left(\omega_1\right)^2 r$$

Where, ω_1 is the speed of rim when engagement starts.

- Centrifugal force acting on each shoe at running speed p = m ω^2r
- Where, $\omega = \left(\dfrac{2\pi N}{60}\right)$
- Friction force acting on each shoe = $\mu(P_c - P_s)$
- The direction of force is perpendicular to the radius of the rim polley.
- Frictional torque on each shoe $\mu(P_c - P_s) \times R$
- Total torque transmitted = Number of shoes $\mu(P_c - P_S)R$

  $= n\mu\left(P_c - P_S\right)R$

Arc = Angle (in radian) × Radiual = θR.

Where, area of contact = lb

- Force exerted on each shoe = plb

  $P_c - P_S = lbp.$

Where, I = Contact length of the shoe

b = Width of the shoe,

p = Pressure intensity on shoe,

q = Angle made by shoe at the centre of spider in radian,

R = Contact radius of shoe = inside radius of the rim of the pulley.

Brakes

Mechanical brakes arrest the energy of a machine or object via force, most commonly friction. Most individuals are familiar with automotive brakes, but mechanical brakes are also essential in material handling, manufacturing, and other power transmission applications.

Mechanical brakes function via force delivered to a body in rotary or linear motion, such as an axle, shaft, or wheel, to slow or stop motion. Mechanical brakes are often in an assembly with a mechanical clutch for engaging and disengaging shafts.

Friction-based brakes utilize a coarse and rugged material (i.e. brake liner) that is tightened or pressed against a body in motion to decelerate. Friction-based braking generates immense heat and some noise, degrading all of the engaged surface areas. Brake capacity decreases with every cycle and requires inspection and replacement. Friction-type brakes are heavily used in automotive applications.

Toothed brakes have tooth-shaped contact surfaces that transmit power without slippage or heat generation. Teeth are engaged only when stopped or running at slow speeds (<20 rpm).

High-performance ceramic disc brake.

Non-contact brakes use a technology such as a magnetic field or eddy currents to provide the braking action. Braking force is generated proportionately to velocity. An advantage of eddy current and magnetic brakes is that there is no friction, so there are no parts that wear out. A disadvantage of these brakes is that they do not have any holding force for stationary objects.

A wrap-spring clutch and brake works by connecting an input shaft and output shaft with an interference fit helical clutch spring. When the input shaft rotates in the direction of the wind of the clutch spring, torque is transmitted to the output shaft. A separate brake spring spins until the brake spring control tang is locked, which tightens the brake spring around the output shaft in position to a stationary brake hub that is bolted to a plate. Simultaneously the clutch spring releases. Wrap-spring brakes provide precise stopping (± 0.5°) in industrial machinery, but fully-mechanical wrap-spring clutches and drives are limited to smaller sizes; larger sizes are electromagnetic.

Wrap-spring clutch and brake

Types

The following brakes are types of friction brakes:

In drum brakes the brake lining is adhered to the external surface of a curved bracket, called a shoe. The most common configuration includes two shoes mounted inside of a drum on a plate. A cylinder presses the shoes onto the insides of the drum to initiate deceleration. A drum brake that presses on the outside of the drum is called a clasp brake; a double clasp brake applies braking pressure to both the inside and outside of the drum.

Drum brake

Cone brakes are a type of drum brake where the drum and shoe are mating sections of conical frustums. The shoe (i.e. cone) is outfitted with brake lining and pressed into the drum (i.e. cup) to apply friction. The advantage is increased surface area and quicker deceleration.

Disc brakes utilize a metal disc, also called a rotor that is connected to the axle. The rotor spins between a caliper that contains between one and 12 cylinders, which pushes a lining material outfitted on a brake pad against the rotor surface.

An Ausco-Lambert disc brake utilizes two discs along with two 360° brake pads positioned inside the discs; one of the pads is stationary. Instead of pinching, the brake pads expand and initial contact between brake lining and disc is provided by actuators. Ball bearings are positioned between the brake pads in tapered recesses. When the pads expand the bearings ride the incline and jam the pads against each other, thereby providing additional, self-sustaining friction to the discs. When stopped the brakes auto-release. The Ausco-Lambert provides space savings, last longer, and run cooler, but are considered oversensitive. They have not been used on autos in 60 years.

Band brakes tighten a ribbon of high-friction material around a pulley attached to the rotating axle; they are often employed on bicycles. If the pull on the band is in the direction of axle rotation the brake is self-energizing. Differential band brakes attach both ends of the brake ribbon to the lever to supply braking power for bi-directional shafts.

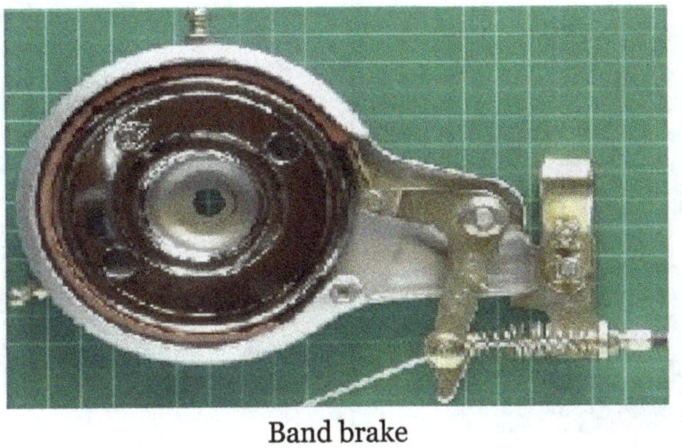

Band brake

Size and Weight

Size and weight are important considerations for mechanical brakes. They add weight to operations, require dedicated systems, and may need a power supply, but brakes are also overengineered to handle the maximum kinetic force that may be applied. Weight also adds to the inertia and dynamic torque experienced by wrap-spring brakes.

Common size considerations for friction brakes include:

- Disc/drum thickness - Width of the friction surface engaged by the pad or shoe.
- Disc/drum diameter - The cross-sectional distance of the disc or drum.
- Total pad area - the square surface area of the brake pad or shoe.
- Wearable friction area - The square surface area of the brake lining.
- Bore size - the inner diameter for the disc or drum.
- Bolt holes - The quantity and size for the bore holes used for axle attachment.
- Running clearance - Distance between brake lining and surface area when brakes are not engaged.
- Lining thickness - The height of brake lining material.

Wrap-Spring Brake Geometries:

- Bore diameter - The inner diameter of the wrap-spring assembly.
- Frame size - The nominative envelope size of the clutch and brake, typically between 2 and 45.

- Diameter - The width across the cross-section of the spring assembly.
- Number of stops - The quantity and degree intervals of stopping locations on the brake collar.
- Thread engagement - Depth of threaded bores for hardware on pulleys/hubs.

Components

Common parts of applicable mechanical brakes are as illustrated.

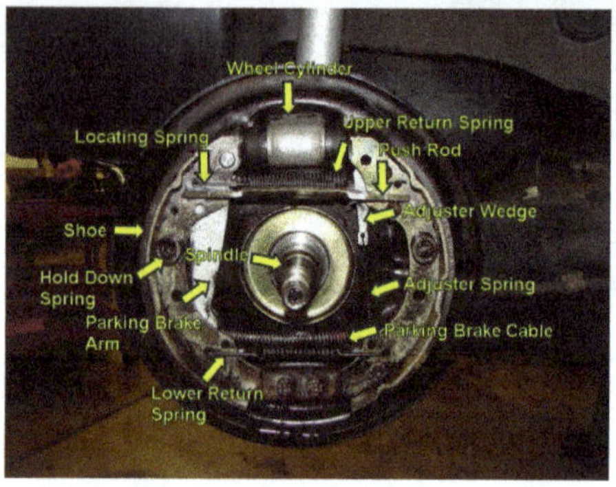

Drum brake

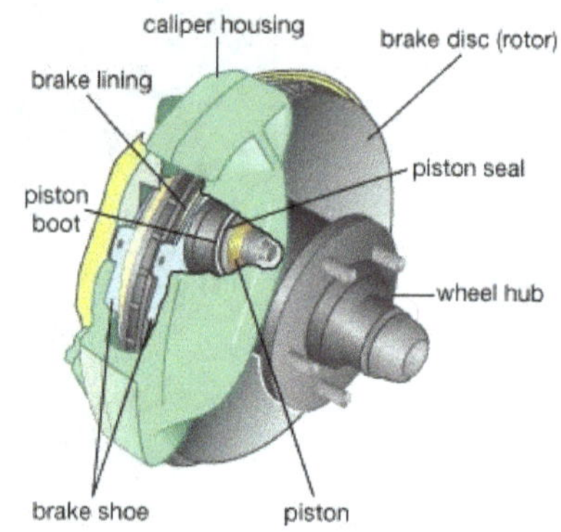

Disc brake

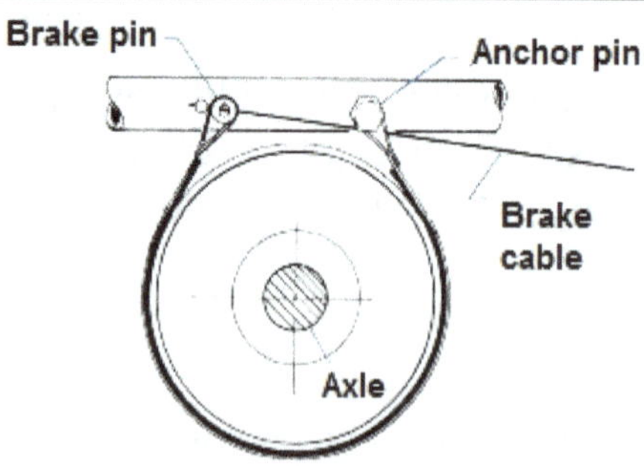

Brand brake

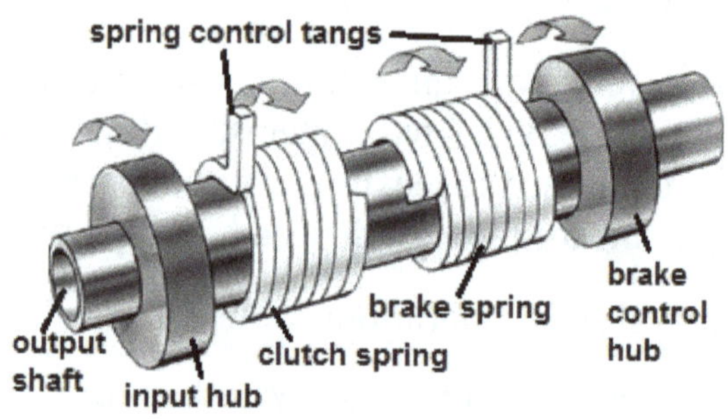

Wrap-spring clutch/brake

Engagement

Friction Brakes

Significant force is required to decelerate components from friction. Brakes that rely solely on manual power supplied by the operator are the purest form of mechanical brake and many need to be self-energizing to be effective.

Dual-function brakes are those that utilize two sources for actuator power, such as pneumatic-mechanical or hydraulic-mechanical. For example, in passenger automobiles the operator depresses a pedal that compresses brake fluid in a linked master cylinder. The master cylinder converts this into hydraulic pressure that actuates the slave cylinders that ultimately provide the braking force. A vacuum servo often helps the driver obtain the foot pressure necessary to pressurize the hydraulic system. Similarly, large commercial vehicles often use pneumatic-mechanical braking systems (i.e. air brakes).

Lever force is the total available braking power from an individual system. This force must be sufficient to decelerate the shaft within a specified time (i.e. braking force).

Braking Torque Calculations

Disc

$$T = WK^2N/308t$$

Where,

T : torque (ft.-lb)

W : weight of rotating member (lb.)

K : radius of gyration of rotating member (ft.)

N : speed of rotating shaft (RPM)

t : stopping time required (seconds).

Drum

$$T = T_T = T_L$$

$$\frac{K \quad Far}{(b \quad K \quad c)}$$

$$\frac{K \quad Far}{(b \quad K \quad c)}$$

$$\frac{4\sin}{(2\theta \quad \sin 2\theta)}$$

Where,

T: torque (ft.-lb.)

T_T: torque for leading shoe

T_L: torque for leading shoe

K: function of the angle of contact

μ : coefficient of friction for brake lining

F: applied force

a: distance between shoe anchor and wheel cylinder

r: drum radius

b: distance between shaft and shoe pins

c: distance from anchor pin to fulcrum

θ: angle of shoe engagement.

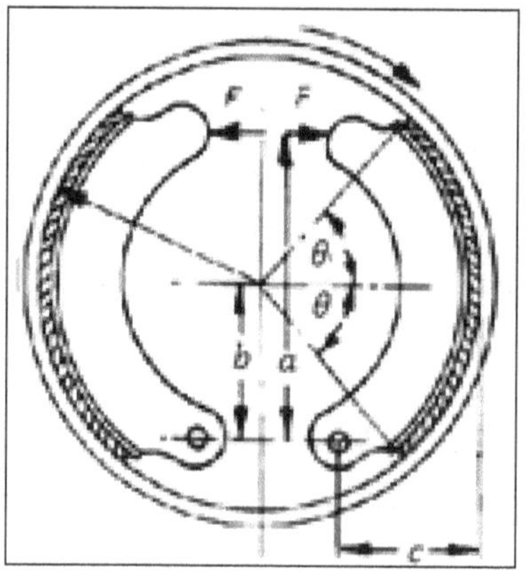

Band

$$T = r(F_1 - F_2)$$
$$F_1 = brp$$
$$F_2 = F\frac{a}{c}$$

Where,

T: torque (ft.-lb.)

r: drum/ pulley radius

F: applied force

b: belt width

p: maximum pressure on friction material

a: lever arm

c: distance from belt attachment of fulcrum.

Wrap-Spring Brakes

In order to accurately start and stop a load, the shaft needs enough torque to engage the clutch spring, as well as enough inertia to engage the brake spring. Calculate the inertia (WR^2) of all rotating components, including shafts, drums and pulley. A reference table will contain data on the inertia of steel shafting per inch of length of thickness, as well as conversion data for non-steel materials. To determine the inertia of hollow parts, first find the inertia of an equivalent solid part (outer diameter), then find the inertia of the hollow area as if it were a shaft (inner diameter). Subtract the inner diameter inertia from the outer diameter inertia, and multiply the per inch value by the length or thickness of the part.

Wrap-spring Braking Calculations

Torque required to engage output shaft	Minimum inertia to engage brake spring
$T = \frac{WR^2 \times s}{3700 \times t} - T_f$ Where: T: torque required to engage output shaft (in.-lb.) WR^2 : load inertia (lb.-in²) S : shaft speed (RPM) t : time to disengage (seconds) T_f: torque to overcome static friction (in.-lb.)	$I = \frac{[(t)(T+T_o)(3700)]}{S} - I_C$ Where: I : minimum inertia needed to engage brakes spring and release clutch spring (lb.-in²) t : time (seconds) T : torque required to engage output shaft (in.-lb.) T_o : drag torque (in.-lb.) S : shaft speed (RPM) I_n : inertia of the clutch/brake output (in.-lb.²)

If the final value for the minimum inertia needed to engage the brakes spring is equal or less than 0, the brake spring has enough inertia to halt the shaft to the determined accuracy. Stop collars engage the brake spring and have one to 20 or more stop positions.

Thermal Management

The friction between brake pads and drum or disc surfaces translates kinetic energy into heat energy, a major concern to retain brake operability. Overheated brakes will transfer excess thermal energy to brake pads, which begin to deteriorate and thereby hasten brake fatigue. Serious overheating causes brake fluid vaporization that can lead to complete failure for disc brake systems. In drum brakes, the drum absorbs up to 95% of all heat generated, so shoes and linings do not experience the same extreme temperatures and brake fluid remains insulated against vaporization. Drum drilling is a technique that allows heat to escape holes created in the drum working surface.

Most disc brakes can exhaust about 3 BTU per hour per square foot of disc surface at 300° F, and higher speed discs can even exhaust up to 5 BTU per hour per square foot. Airflow from vehicle locomotion is often enough to cool disc brakes on vehicles. Discs that experience extreme temperatures or are on stationary equipment are subject to both passive (increased surface area, ventilation slots, multiple calipers, materials changes) and active (compressed air) solutions.

Band brakes have poor cooling capacities and suffer increased brake fade as a result.

Disengagement

Complete release of mechanical brakes results in brake drag that shortens braking components service life and also wastes energy. For many friction brakes, a spring holds the brake in the disengaged position until a force overcomes the spring preload. Brakes can also be supplied as fail-safe brakes, or brakes where the spring keeps the brake engaged and power is needed to remove the spring's braking force.

For band brakes, disengaging the brake requires loosening the brake band, which can be done automatically when the shaft is stopped (as with self-energizing brakes) or by releasing the brake lever or other activation mechanism. The brake spring in wrap-spring clutches and brakes automatically releases when the shaft has stopped.

Materials

Brake lining materials vary considerably by application. Materials must be soft, tough, heat resistant, and possess a high coefficient of dynamic friction. They are applied to pads or shoes made of welded or riveted sheet metal.

- Non-metallic: Composite organic or synthetic materials, including cellulose, aramids, polyacrylnitrile, and sintered glass. Non-metallic linings minimize rotor wear, but have short service lives.
- Semi-metallic: Composite materials filled with metallic flakes to improve wear resistance and service life with increased wear on the rotor or drum; requires higher braking torque.
- Metallic: Often reserved for performance or high energy applications; composed of sintered steel that wears quickly on rotors, requires higher braking torque, and generates noise.
- Ceramic: Clay and porcelain mixed with copper flakes and filaments. Moderate durability, lifespan, and torque requirements with zero perceived sound. However high operating temperatures can warp pads and other components.

Calipers are typically metal plates with cylinders made of plastic, aluminum, or chrome-plated steel. Disc materials include:

- Ceramic composite: Carbon fiber-reinforced ceramic provides stable friction at high speed and all temperatures, is 50% lighter than grey iron, but is also quite expensive.
- Grey iron: Cast iron with graphitic microstructure that has high heat and damping capacities with wear resistance and machinability. It is the most common material for automobile rotors.
- Steel: Stainless steel is commonly integrated into bicycle and motorcycle brakes, however has poor thermal conductivity. Steel may be acceptable for applications with higher thermal requirements, but will also corrode.
- Aluminum: A lightweight metal with excellent thermal characteristics; only suitable for low RPM applications because of decreased wear resistance and strength; sometimes found on bicycles.
- Titanium: Only compatible with organic or resin-based liners; lightweight with good strength and corrosion resistance, but has low surface friction and lifespan.

Some common drum materials are:

- Cast iron: The most common brake drum material.
- Aluminum: Lighter than iron with improved heat dissipation and brake fade; needs iron or steel liner to retain structural integrity.

- Steel: Steel shells are a common means to reduce weight and improve cast iron drum performance.
- Composite: Hybrid materials, such as cast-aluminum filled with silicon carbide, provides weight savings with adequate friction and wear resistance.

Couplings

A coupling is a mechanical device which is used to hold two rotating shafts together. It is a permanent joint, unlike clutch which can be disengaged on the will of operator. With the help of couplings, we can join two shafts which are intersecting, colinear or parallel (with a small distance). Oldham coupling is used to connect two parallel shafts which are a small distance apart.

Difference between Clutch and Coupling

We all know that a clutch (which is generally used in automobiles) is also used for connecting two rotating shafts but there is a basic difference between a clutch and coupling. In clutch we can disengage the rotating shaft at our will whenever we want but in coupling it is not possible.

Practical Example of Coupling

In our domestic electricity generator, we see that a shaft is rotated by an engine by converting chemical energy of fuel into mechanical energy. This rotating shaft of engine is then connected with the shaft of an electric alternator with the help of a coupling. And finally, the electric alternator converts mechanical energy of shaft into electric energy. Here coupling is an important part which transmits rotating motion of engine shaft to electric alternator shaft.

Requirements of a Good Coupling

- It should be capable of transmitting torque from driving shaft to drive shaft.
- It should align the shafts properly.
- It should be able to dismantle for the purpose of maintenance.
- It should provide safety to man and machine in case of coupling failure.

Types of Couplings

There are two types of couplings:

- Rigid coupling,
- Flexible coupling.

A flexible coupling can accommodate a slight misalignment between shafts (up to 0.5-degree angular misalignment and 5 mm axial misalignment) but a rigid coupling cannot tolerate any such misalignment. Examples of rigid coupling muff coupling, clamp coupling, rigid flange coupling etc. Example of flexible coupling is Bushed pin flexible coupling.

Difference between Rigid and Flexible Couplings

- Flexible coupling can accommodate a slight misalignment between shafts but a rigid coupling cannot accommodate any misalignment.
- Flexible coupling can absorb shocks and vibrations but a rigid coupling cannot absorb shocks and vibrations.
- Rigid coupling is simple and inexpensive but flexible couplings are expensive.

Muff Coupling

A muff coupling is also known as a sleeve or box coupling. It is a type of rigid coupling. It consists a sleeve and a key which is used to join input and output shafts.

Advantages of Muff Coupling

- It is simple it has only two parts a sleeve and a key.
- Since it has no projecting parts hence it is safe to use.
- It has compact construction.
- It is cheaper compared to other types of couplings.

Disadvantages of Muff Coupling

- It is difficult to assemble or dismantle.
- Since it is a rigid coupling so it cannot accommodate any misalignment.
- Due to absence of flexible elements it cannot absorbs shocks and vibrations.
- It requires more axial space compared with other types of couplings.

Machine Tool

Machine tool refers to any stationary power-driven machine that is used to shape or form parts made of metal or other materials. The shaping is accomplished in four general ways: (1) by cutting excess material in the form of chips from the part; (2) by shearing the material; (3) by squeezing metallic parts to the desired shape; and (4) by applying electricity, ultrasound, or corrosive chemicals to the material. The fourth category covers modern machine tools and processes for machining ultrahard metals not machinable by older methods.

Machine tools that form parts by removing metal chips from a workpiece include lathes, shapers and planers, drilling machines, milling machines, grinders, and power saws. The cold forming of metal parts, such as cooking utensils, automobile bodies, and similar items, is done on punch presses, while the hot forming of white-hot blanks into appropriately shaped dies is done on forging presses.

Modern machine tools cut or form parts to tolerances of plus or minus one ten-thousandth of an inch (0.0025 millimetre). In special applications, precision lapping machines can produce parts that are within plus or minus two millionths of an inch (0.00005 millimetre). Because of the precise dimensional requirements of the parts and the heavy cutting forces exerted on the cutting tool, machine tools combine weight and rigidity with delicate accuracy.

Machine-tool Characteristics

All machine tools must provide work-holding and tool-holding devices and means for accurately controlling the depth of the cut. The relative motion between the cutting edge of the tool and the work is called the cutting speed; the speed in which uncut material is brought into contact with the tool is called the feed motion. Means must be provided for varying both.

Because an overheated tool may lose its cutting ability, temperatures must be controlled. The amount of heat that is generated depends on the shearing force and the cutting speed. Because the shearing force varies with the material being cut and the tool material varies in its tolerance for high temperatures, the optimum cutting speed depends both on the material being cut and the cutting-tool material. It is also influenced by the rigidity of the machine, the shape of the workpiece, and the depth of the cut.

Cutting Tools

Metal-cutting tools are classified as single point or multiple point. A single-point cutting tool can be used for increasing the size of holes, or boring. Turning and boring are performed on lathes and boring mills. Multiple-point cutting tools have two or more cutting edges and include milling cutters, drills, and broaches.

There are two types of operation; either the tool is moving on a straight path against the stationary workpiece, as on a shaper, or the workpiece is moving against the stationary tool, as on a planer. Relief or clearance angles must be provided to prevent the tool surface below the cutting edge from rubbing against the workpiece. Rake angles are often provided on cutting tools to cause a wedging action in the formation of chips and to reduce friction and heat.

Tool Materials

In order to remove chips from a workpiece, a cutting tool must be harder than the workpiece and must maintain a cutting edge at the temperature produced by the friction of the cutting action.

Carbon Steel

Steel with a carbon content ranging from 1 to 1.2 percent was the earliest material used in machine tools. Tools made of this carbon steel are comparatively inexpensive but tend to lose cutting ability at temperatures at about 400° F (205 °C).

High-speed Steel

In 1900 the introduction of high-speed steel permitted the operation of tools at twice or three times the speeds allowable with carbon steel, thus doubling or trebling the capacities of the world's machine shops. One of the most common types of high-speed steel contains 18 percent tungsten, 4 percent chromium, 1 percent vanadium, and only 0.5 to 0.8 percent carbon.

Cast Alloys

A number of cast-alloy cutting-tool materials have been developed; these nonferrous alloys contain cobalt, chromium, and tungsten and are particularly effective in penetrating the hard skin on cast iron and retaining their cutting ability even when red hot.

Cemented Tungsten Carbide

This material was first used for metal cutting in Germany in 1926. Its principal ingredient is finely divided tungsten carbide held in a binder of cobalt; its hardness approaches that of a diamond. Tungsten carbide tools can be operated at cutting speeds many times higher than those used with high-speed steel.

Oxides

Ceramic, or oxide, tool tips are one of the newest developments in cutting-tool materials. They consist primarily of fine aluminum oxide grains, which are bonded together.

Diamonds

Diamonds have been used for many years for truing grinding wheels, in wire-drawing dies, and as cutting tools. For cutting applications they are used largely for taking light finishing cuts at high speed on hard or abrasive materials and for finish-boring bronze and babbitt-metal bearings.

Cutting Fluids

In many machine-tool operations, cutting fluids or coolants are used to modify the harmful effects of friction and high temperatures. In general, the major functions of a coolant are to lubricate and cool. When cutting a screw thread, either on a lathe or with a tap, the lubricating function is most important; in production-grinding operations, the cooling function predominates. Water is an excellent cooling medium, but it corrodes ferrous materials. Lard oil has excellent lubricating qualities, but it tends to become rancid. Sulfurized mineral oil is one of the most popular coolants. The sulfur tends to prevent chips from the work from welding to the tip of the tool. For sawing and grinding operations, soluble oil, which is an oily emulsion freely miscible in water, is commonly used.

Basic Machine Tools

Hundreds of varieties of metal machine tools, ranging in size from small machines mounted on workbenches to huge production machines weighing several hundred tons, are used in modern industry. They retain the basic characteristics of their 19th and early 20th-century ancestors and are still classed as one of the following: (1) turning machines (lathes and boring mills), (2) shapers and planers, (3) drilling machines, (4) milling machines, (5) grinding machines, (6) power saws, and (7) presses.

Turning Machines

The engine lathe, as the horizontal metal-turning machine is commonly called, is the most important of all the machine tools. It is usually considered the father of all other machine tools because many of its fundamental mechanical elements are incorporated into the design of other machine tools.

Metal being cut on a lathe.

The engine lathe is a basic machine tool that can be used for a variety of turning, facing, and drilling operations. It uses a single-point cutting tool for turning and boring. Turning operations involve cutting excess metal, in the form of chips, from the external diameter of a workpiece and include turning straight or tapered cylindrical shapes, grooves, shoulders, and screw threads and facing flat surfaces on the ends of cylindrical parts. Internal cylindrical operations include most of the common hole-machining operations, such as drilling, boring, reaming, counterboring, countersinking, and threading with a single-point tool or tap.

Boring involves enlarging and finishing a hole that has been cored or drilled. Bored holes are more accurate in roundness, concentricity, and parallelism than drilled holes. A hole is bored with a single-point cutting tool that feeds along the inside of the workpiece. Boring mills have circular horizontal tables that rotate about a vertical axis, and they are designed for boring and turning operations on parts that are too large to be mounted on a lathe.

Shapers and Planers

Shaping and planing operations involve the machining of flat surfaces, grooves, shoulders, T-slots, and angular surfaces with single-point tools. The largest shapers have a 36-inch cutting stroke and can machine parts up to 36 inches long. The cutting tool on the shaper oscillates, cutting on the forward stroke, with the workpiece feeding automatically toward the tool during each return stroke.

Planing machines perform the same operations as shapers but can machine longer workpieces. Some planer can machine parts up to 50 feet long. The workpiece is mounted on a reciprocating table that moves the workpiece beneath a cutting tool. This tool, which remains stationary during the cutting stroke, automatically feeds into the workpiece after each cutting stroke.

Drilling Machines

Drilling machines, also called drill presses, cut holes in metal with a twist drill. They also use a variety of other cutting tools to perform the following basic hole-machining operations: (1) reaming, (2) boring, (3) counterboring, (4) countersinking, and (5) tapping internal threads with the use of a tapping attachment.

Milling Machines

A milling machine cuts metal as the workpiece is fed against a rotating cutting tool called a milling cutter. Cutters of many shapes and sizes are available for a wide variety of milling operations. Milling machines cut flat surfaces, grooves, shoulders, inclined surfaces, dovetails, and T-slots. Various form-tooth cutters are used for cutting concave forms and convex grooves, for rounding corners, and for cutting gear teeth.

Milling machines are available in a variety of designs that can be classified as the following: (1) standard knee-and-column machines, including the horizontal and the vertical types; (2) bed-type or manufacturing machines; and (3) machines designed for special milling jobs.

Grinding Machines

Grinding machines remove small chips from metal parts that are brought into contact with a rotating abrasive wheel called a grinding wheel or an abrasive belt. Grinding is the most accurate of all of the basic machining processes. Modern grinding machines grind hard or soft parts to tolerances of plus or minus 0.0001 inch (0.0025 millimetre).

The common types of grinding machines include the following: (1) plain cylindrical, (2) internal cylindrical, (3) centreless, (4) surface, (5) off-hand, (6) special, and (7) abrasive-belt.

Power Saws

Metal-cutting power saws are of three basic types: (1) power hacksaws, (2) band saws, and (3) circular disk saws. Vertical band saws are used for cutting shapes in metal plate, for internal and external contours, and for angular cuts.

Presses

This large class of machines includes equipment used for forming metal parts by applying the following processes: shearing, blanking, forming, drawing, bending, forging, coining, upsetting, flanging, squeezing, and hammering. All of these processes require presses with a movable ram that can be pressed against an anvil or base. The movable ram may be powered by gravity, mechanical linkages, or hydraulic or pneumatic systems.

Appropriate die sets, with one part mounted on the movable ram and the matching part mounted on the fixed bed or platen, are an integral part of the machine. Punch presses stamp out metal parts from sheet metal and form the parts to the desired shape. Dies with cavities having a variety of shapes are used on forging presses that form white-hot metal blanks to the desired shapes. Power presses are also used for shearing, bending, flanging, and shaping sheet metal parts of all sizes. Power presses are made in various sizes, ranging from small presses that can be mounted on a workbench to machines weighing more than 1,000,000 pounds (450,000 kilograms).

References

- Machine-design-5: chegg.com, Retrieved 4 May, 2019
- Machine-design: myindialist.com, Retrieved 11 July, 2019
- Manufacturing-process-meaning-and-types: engineeringarticles.org, Retrieved 17 February, 2019
- Types-of-casting-processes, custom-manufacturing-fabricating: thomasnet.com, Retrieved 20 January, 2019
- Manufacturing-process-meaning-and-types: engineeringarticles.org, Retrieved 1 March, 2019
- Types-machining, custom-manufacturing-fabricating: thomasnet.com, Retrieved 11 February, 2019
- Manufacturing-process-meaning-and-types: engineeringarticles.org, Retrieved 30 April, 2019
- Mechanical-brakes, clutches-brakes, motion-controls: globalspec.com, Retrieved 20 June, 2019
- Machine-tool: britannica.com, Retrieved 25 August, 2019

Permissions

All chapters in this book are published with permission under the Creative Commons Attribution Share Alike License or equivalent. Every chapter published in this book has been scrutinized by our experts. Their significance has been extensively debated. The topics covered herein carry significant information for a comprehensive understanding. They may even be implemented as practical applications or may be referred to as a beginning point for further studies.

We would like to thank the editorial team for lending their expertise to make the book truly unique. They have played a crucial role in the development of this book. Without their invaluable contributions this book wouldn't have been possible. They have made vital efforts to compile up to date information on the varied aspects of this subject to make this book a valuable addition to the collection of many professionals and students.

This book was conceptualized with the vision of imparting up-to-date and integrated information in this field. To ensure the same, a matchless editorial board was set up. Every individual on the board went through rigorous rounds of assessment to prove their worth. After which they invested a large part of their time researching and compiling the most relevant data for our readers.

The editorial board has been involved in producing this book since its inception. They have spent rigorous hours researching and exploring the diverse topics which have resulted in the successful publishing of this book. They have passed on their knowledge of decades through this book. To expedite this challenging task, the publisher supported the team at every step. A small team of assistant editors was also appointed to further simplify the editing procedure and attain best results for the readers.

Apart from the editorial board, the designing team has also invested a significant amount of their time in understanding the subject and creating the most relevant covers. They scrutinized every image to scout for the most suitable representation of the subject and create an appropriate cover for the book.

The publishing team has been an ardent support to the editorial, designing and production team. Their endless efforts to recruit the best for this project, has resulted in the accomplishment of this book. They are a veteran in the field of academics and their pool of knowledge is as vast as their experience in printing. Their expertise and guidance has proved useful at every step. Their uncompromising quality standards have made this book an exceptional effort. Their encouragement from time to time has been an inspiration for everyone.

The publisher and the editorial board hope that this book will prove to be a valuable piece of knowledge for students, practitioners and scholars across the globe.

Index

www.ingramcontent.com/pod-product-compliance
Lightning Source LLC
LaVergne TN
LVHW080913210926
848485LV00004B/645